VLSI
DESIGN

VLSI CIRCUITS SERIES

Series Editor

Wai-Kai Chen

University of Illinois at Chicago

VLSI
DESIGN

M. Michael Vai, Ph.D.

CRC Press
Boca Raton London New York Washington, D.C.

Library of Congress Cataloging-in-Publication Data

Vai, M. Michael.
    VLSI design / M. Michael Vai.
        p. cm.
    Includes bibliographical references and index.
    ISBN 0-8493-1876-9 (alk. paper)
    1. Intergrated circuits—Very large scale integration—Design and construction. I. Title.

TK7874.75.V25 2000
621.39′5—dc21
                                              00-059896
                                                                   CIP

Visit the CRC Press Web site at www.crcpress.com

© 2001 by CRC Press LLC

No claim to original U.S. Government works
International Standard Book Number 0-8493-1876-9
Library of Congress Card Number 00-059896
Printed in the United States of America   4 5 6 7 8 9 0
Printed on acid-free paper

Preface

Officially, VLSI is the acronym of *very large scale integration*. When I was a graduate student in the early 1980s, unofficially VLSI was the acronym for *very large salary income*. When my teaching career began at Northeastern University, Boston, Massachusetts, in the late 1980s, my students complained that VLSI was *very long student involvement*.

Two decades have passed since Carver Mead and Lynn Conway published their classic book *Introduction to VLSI Systems*. Many excellent books have been written for the subject of VLSI design. I have used a number of them in my undergraduate and graduate VLSI design courses. These books apparently have served the purpose of introducing VLSI design techniques to my electrical/computer engineering students. However, in this period of time, VLSI design has evolved from full-custom layouts to IP (intellectual property) cores and SOC (silicon on a chip) designs. I have been longing for a book that would provide a *complete* overview of the VLSI technologies. This book was written to quench such a desire.

This book is written to provide a complete and congregated view of VLSI engineering, which ranges from CMOS logic design to physical design automation to fault tolerant array processors. It is my hope that this book can provide the reader with the foundation to pursue answers to the important questions below:

- What is VLSI?
- How to VLSI?
- When to VLSI?
- What to VLSI?

Chapters 1 to 7 describe, from a digital circuit engineer's view, the basic CMOS design techniques. Physical pictures about CMOS circuits such as the benefits and limitations of using simplified models are emphasized. Chapter 8 discusses the top-down design process of two projects, a simple microprocessor and a field programmable gate array (FPGA). These projects provide opportunities for students to put together the knowledge that they have learned from this course as well as from other computer engineering/science courses. They also provide an ideal platform to explore the important tradeoffs and compromises facing VLSI designers every day.

Chapter 9 introduces the testing of VLSI circuits. This is followed by an overview of the working principles of physical design automation tools in Chapter 10. CAD tools have become an indispensable part of VLSI design process. An understanding of the strength and weakness of available CAD tools is an important factor in creating high performance circuits.

The remaining chapters of the book build on the foundation laid by the first nine chapters and concentrate on providing answers to the last two questions. The message that VLSI is a tool to develop innovative algorithms and architectures to solve problems that are previously intractable is conveyed. Chapter 11 presents parallel processing in a more conventional view. The design and application of VLSI array processors are discussed in Chapter 12 and Chapter 13.

VLSI design is a subject that involves a broad range of knowledge. Every chapter in this book can be expanded into a full-blown reference book on the subject it discusses. Only then, it would not be able to conveniently serve the purpose of a textbook. In order to maintain a depth while broadening its width, I have used individual term projects to complement class discussion.

After reading Chapter 1, every student selects one topic in VLSI design to perform an independent term project. For students who want to pursue a research thesis in VLSI Engineering, they can use this opportunity to initiate their research projects. Students can identify specific areas in VLSI to study and research along the structured lectures. This teaching approach has served my students extremely well in the last decade, regardless of their plans and goals (standard cell designers, ASIC designers, system engineers, application engineers, CAD tool developers, academic and industrial researchers, etc.).

I have suggested to my students the following steps in pursuing this individual term project. First, perform a literature research. A number of VLSI related journals, magazines, conference proceedings, and web-sites are listed at the end of Chapter 1. In the second step, identify a term project topic. Three types of term projects have been pursued. The first type is a design project. The design and verification of a simple microprocessor have been chosen by some students (see Chapter 8). The design can actually be started once Chapter 3 is covered so that it can be taped out to MOSIS (a brokerage service for silicon foundries) before the end of the semester.

The second type of project involves the programming of CAD tools. Tools such as placement and routing have been popular candidates for this type of project. The third type of project is to study a few technical publications on a VLSI design topic and write a literature research report.

The authoring of a book is a big project. This is especially true in my case after I left my teaching position at Northeastern University to work for MIT Lincoln Laboratory in the Summer of 1999. The understanding and encouragement of the CRC Press staff have been indispensable to the completion of this book. I am indebted to numerous VLSI experts in this field; their knowledge and techniques have been drawn on heavily to form this book. I would also like to thank the thousands of students who had taken my courses, especially those (about a hundred) who have used its draft copy in doing so. Last, but not least, the strong support of my wife, Cecily, and my son, Alex, is truly appreciated.

M. Michael Vai, Ph. D.
Lexington, Massachusetts
Summer, 2000

About the Author

M. Michael Vai received his B.S. degree from National Taiwan University, Taipei, Taiwan, in 1979, and his M.S. and Ph.D. degrees from Michigan State University, East Lansing, Michigan, in 1985 and 1987, respectively, all in electrical engineering.

Until July 1999, he was on the faculty of Electrical and Computer Engineering Department, Northeastern University, Boston, Massachusetts. At Northeastern University, he developed and taught the graduate and undergraduate VLSI courses. In May 1999, he received the Outstanding Professor Award presented by the students of Electrical and Computer Engineering, Northeastern University. Currently, he is a technical staff member at Lincoln Laboratory, Massachusetts Institute of Technology, Lexington, Massachusetts.

He has worked and published extensively in VLSI, microelectronics, computer engineering, and engineering education.

Contents

Chapter 1 Introduction

The story begins ...

Our story began in 1948 when William Shockley, John Bardeen, and Walter H. Brattain of Bell Laboratories invented the transistor. Currently, a majority of integrated circuits are built with a type of transistor called the metal-oxide-semiconductor field-effect transistor (MOSFET).

An integrated circuit (IC) is a tiny semiconductor chip, on which a complex of electronic components and their interconnections are fabricated with a set of pattern defining masks.[1] The technique of fabricating an IC with a set of masks is somewhat analogous to that of creating a photograph with a negative. We describe a typical IC fabrication process in Chapter 3.

The concept of fabricating multiple electronic devices and their interconnections entirely on a small semiconductor chip was proposed shortly after the transistor was invented. ICs began to become commercially available in the early 1960s. IC design and fabrication technologies have been developed at a phenomenal rate since then.

The evolution of IC technology is measured by the number of components integrated on a single chip. The IC industry has gone through the milestones of small scale integration (SSI), medium scale integration (MSI), large scale integration (LSI), and very large scale integration (VLSI).

Fig. 1.1 shows the approximate number of transistors available on a single chip in each of these periods.

Integration Scale	Number of Components	Examples
SSI	< 10 transistors	Logic gates
MSI	10 - 1,000 transistors	Adders, counters
LSI	1,000 - 10,000 transistors	Multipliers
VLSI	> 10,000 transistors	Microprocessors

Fig. 1.1 Integration scales.

In 1965, Gordon Moore (a cofounder of Intel) made an observation and prediction on the development of semiconductor technology. His prediction is that the effectiveness of semiconductor technology, as roughly measured by the maximum number of transistors on a chip, approximately doubles every eighteen months. This forecast, now commonly known as "Moore's law," has been accurate for the last four decades. A depiction of Moore's law is shown in Fig. 1.2. Note that the y-axis is

---

[1] Both terms "integrated circuit" and "chip" will be used interchangeably in this book.

drawn in a logarithmic scale. The rapid advance of IC technology is evident from the fact that the on-chip transistor count has increased one million times in three decades.

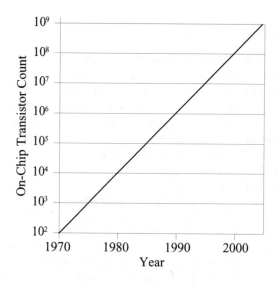

Fig. 1.2 Moore's law.

The exponential growth of IC technology is also evident in the smallest manufacturable feature size (minimum line width), which has been reduced from around 10 μm in the 1970s to less than 0.2 μm in 2000 and continues to decrease. One micrometer (μm), also called a micron, equals 1×10^{-6} m and is commonly used as the basic length unit in IC design. The use of an even smaller length unit, the nanometer (nm), is getting increasingly popular. One μm equals 1000 nm. Physicists predict that the lower bound of feature size is around 0.05 μm (50 nm) for the current form of MOSFET ICs.

While Fig. 1.1 shows that any circuits with more than 10,000 transistors are considered to be in the VLSI category, currently millions of transistors are routinely integrated into a single chip. For example, the Intel Pentium III processor has 28 million transistors and is fabricated with a technology that provides a minimum feature size of 0.18 μm. It operates at 700 MHz and performs 2000 million instructions per second (MIPS).

The creation of a VLSI circuit involves a large number of activities such as logic design, modeling, simulation, testing, and fabrication, each of which can in turn be divided into many tasks. A broad spectrum of knowledge ranging from solid state physics to logic circuit design is required to successfully carry out all these steps. The rapid advancement of manufacturing capability has apparently enabled the growth of integration density. However, without the methodology specifically developed for the design of VLSI circuits, the potential of advanced manufacturing techniques would not have been realized. We will describe the VLSI design methodology in the next section.

1.1   VLSI Design Methodology

The VLSI design methodology developed by Mead and Conway and others in the late 1970s released the IC designers from the details of semiconductor manufacturing, so they could concentrate their efforts on coping with the circuit functionality.[2] This design methodology, illustrated in Fig. 1.3 as an exchange of information between the IC designers (design team) and semiconductor engineers (silicon foundry), is largely responsible for the success of the IC industry in the last two decades. The third participant of the IC industry is the computer-aided design (CAD) tool providers who develop software tools to support design activities. Fig. 1.3 also shows that CAD tool providers often optimize their tools for specific fabrication processes.

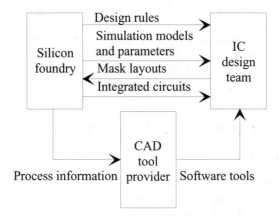

Fig. 1.3 Relationship between a silicon foundry,
an IC design team, and a CAD tool provider.

Fig. 1.3 shows that the design and fabrication of an IC are typically performed by two separate teams with different expertise — a silicon foundry (an IC manufacturer that provides fabrication services to clients) and an IC design team. A silicon foundry develops and provides design rules compatible with its fabrication technology to its clients. Design rules are a set of dimensional rules (e.g., minimum line widths, minimum spacings, etc.) determined by considering the physical limitations of a fabrication process. Furthermore, simulation models of the active and

---

[2] C. A. Mead and L. A. Conway, *Introduction to VLSI Systems*, Addison-Wesley, 1980.

passive devices fabricated by a certain process are developed and made available to the logic designers.

The IC designers use CAD tools to create mask layouts which represent how components and connections should be formed on a semiconductor wafer. These designs, after being simulated and verified, are delivered to the silicon foundry, which produces ICs according to these mask layouts.

The chance of successfully fabricating a chip design that conforms to the design rules is very high. However, the reader should realize that a successfully processed chip is not necessarily equivalent to a functionally correct one. The principle of "garbage in, garbage out" applies. The function and performance of a chip can be predicted by simulation with reasonable accuracy.

The most important design principle emphasized in the VLSI design methodology is to divide-and-conquer. Instead of dealing with an entire VLSI circuit altogether at the same time, the designer partitions it into smaller and thus more manageable parts. These parts may further be broken down into even smaller building blocks. The partitioning of a VLSI system into increasingly smaller subsystems so that they can be handled efficiently is called the hierarchical design methodology. CAD tools have been developed to automate the steps in a hierarchical design.

In this chapter, we provide a general overview of the VLSI design process. It is the intention of the chapter to provide the reader with a qualitative understanding of the steps involved in the VLSI design process. The details of these steps are explained and discussed in later chapters.

1.2   VLSI Design — An Overview

The time will come when we want to discuss the process of designing sophisticated VLSI systems. At this moment, we need a representative, yet relatively simple circuit to demonstrate a typical design process. The design of an arithmetic adder is selected here for this purpose. The reader should notice the similarities and differences between the design of a VLSI chip and a printed circuit board system.

Technology

It is reasonable to assume that a design should begin by studying the characteristics of available fabrication processes. This overview thus begins with an introduction to the IC fabrication technology. However, as explained in the last section, the modern IC design methodology separates the designers from the semiconductor processing tasks so there is no urgent need to select a specific IC fabrication technology at the initial phase of the design. A general knowledge of the fabrication process is sufficient at this point.

The major materials that are used to build semiconductor devices and integrated circuits include silicon (Si), germanium (Ge), and gallium arsenic (GaAs). Silicon is by far the most popular material for VLSI fabrication. All signs indicate that, in the foreseeable future, silicon will continue to be the preferred semiconductor material of IC manufacturers, especially those targeting the digital IC market.

Two types of silicon based device, the bipolar junction transistor (BJT) and the metal-oxide-semiconductor field-effect transistor (MOSFET), have been used to construct logic circuits. The complementary MOSFET (CMOS) technology, which employs both n-channel and p-channel transistors to form logic circuits, has the following advantages over the bipolar technology and dominates digital logic ICs:

- The nature of a CMOS circuit allows it to operate with low power and thus a higher integration density is possible.
- A MOSFET occupies a smaller area than a BJT, which also leads to a higher circuit density.
- A MOSFET has a very high input impedance and can be modeled as a switch. This feature simplifies the design and analysis of CMOS logic circuits.
- CMOS circuits have the largest logic swings and thus excellent noise margins.

Chapter 2 will further discuss the design and analysis of CMOS logic circuits. While the CMOS technology has a number of preferable features, logic circuits based on BJTs, such as transistor-transistor logic (TTL) and emitter-coupled logic (ECL), have the advantages of high speed and large current capability. Both BJTs and MOSFETs can coexist in the same circuit. This technology, called the BiCMOS technology, offers the high speed of BJTs and the low power advantage of CMOS technology. A brief introduction to the BiCMOS technology is presented in Chapter 6.

The CMOS technology requires both n-channel and p-channel MOSFETs to coexist on the same substrate. In order to do this, a portion of the substrate must be converted to accommodate the transistors of the opposite type. The n-well CMOS technology uses a p-type wafer as the substrate in which n-channel transistors can be formed directly. P-channel transistors are formed in an n-well which is created by converting a portion of the p-type wafer from being hole (p-type carrier) rich into electron (n-type carrier) rich. Other possibilities include the p-well CMOS technology and the twin-well CMOS technology. More details on these will be presented in Chapter 3.

Two important properties of a selected CMOS technology are its minimum manufacturable feature size and the number of available routing layers. Typically, the minimum feature size of a CMOS technology determines the shortest transistor channel length possible. For example, the shortest transistor channel length is about 0.25 μm in a 0.25 μm-CMOS technology. Generally speaking, a smaller feature size implies faster transistors and a higher circuit density.

Another important property of a CMOS technology is the number of available routing layers for component interconnections. Routing layers insulated from each other by silicon dioxide (SiO_2) allow a wire to cross over another line without making undesired contacts. Connecting paths called vias can be formed between two routing layers to make inter-layer connections as required. A typical CMOS process has 3 to 7 layers available for the routing purpose.

Behavioral (Functional) Design

The first step in designing a circuit is usually to specify a behavioral (functional) representation of the circuit. This involves a description of how the circuit should communicate with the outside world. Typical issues at this representation level include the number of input-output (I/O) terminals and their relations. A well-defined behavioral description of a circuit has a major benefit. It allows the designer to optimize a design by choosing a circuit from a set of structurally different, yet functionally identical ones that conform to the desired behavioral representation.

Let us use the adder design to contemplate the behavioral issues that the designer has to resolve. One way to appreciate the behavioral representation problem is to imagine the role of a VLSI designer who has received the assignment of designing, say, a "fast, low-power adder." The designer must develop specific technical specifications of this adder. The reader should keep in mind that a VLSI design process is highly iterative and none of the specifications initially selected for the circuit is final until a satisfactory IC is produced.

The input operands of the adder can be 16-bit or 32-bit, unsigned or signed, integer or fraction, fixed point or floating point numbers. Note that these variations are just a few of many possibilities. The choice of a number system apparently has significant effects on the design complexity and performance of the adder. For example, adding two unsigned n-bit integers together produces an $(n + 1)$-bit sum and this observation determines the width of the adder output. Alternatively, the sum can be limited to be an n-bit integer. An "overflow" flag is raised if the result is too large to be represented by n bits. There are apparently many other decisions to be made but we will not get into the details here. A graphical behavioral representation of our example adder is illustrated in Fig. 1.4. We choose to add two 32-bit unsigned integers (a and b) to produce a 33-bit sum (s).

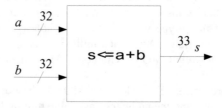

Fig. 1.4 Graphical behavioral representation of an adder.

Instead of the graphical representation shown in Fig. 1.4, the adder can also be represented by a hardware description language model. A hardware description language is a programming language with a syntax specially designed to allow the description of circuits. Currently, the most popular hardware description languages are VHDL (Very high speed integrated circuit Hardware Description Language) and Verilog.

For example, in VHDL, the adder given in Fig. 1.4 may be described as

```
-- Interface
ENTITY adder IS
   PORT (a, b: IN unsigned(0 to 31);
      s: OUT unsigned(0 to 32));
END adder;

-- Behavioral representation
ARCHITECTURE behavioral OF adder IS
BEGIN
   s <= a + b;
END behavioral;
```

It is beyond the scope of this book to explain how to write VHDL models, but the reader with some software programming knowledge should not have much difficulty understanding the above code by considering the following key points.[3] Any line that begins with "--" is a comment line. All VHDL keywords are capitalized in the code examples. The line "s <= a + b;" specifies that the sum of a and b is assigned (<=) to s. The above VHDL code consists of two parts. The first part, consisting of the first five lines, is the "interface" defined for the adder. It specifies that the adder has three ports. Ports a and b are 32-bit unsigned inputs and port s is a 33-bit unsigned output. The second part describes the relationship between the inputs and output of the adder, which simply states that output s is the sum of inputs a and b.

Timing information can also be included in a behavioral model. For example, we can specify that the delay of this adder is, say, 5 nanoseconds (ns). In the VHDL code, an "after" clause may be added to the behavioral description line so that it reads

```
s <= a + b AFTER 5 ns;
```

Structural (Architectural) Design

CAD tools typically allow a circuit to be represented at multiple abstraction levels. Components described at different abstraction levels can be assembled together to form a system. Such a system can be simulated with a mixed-level simulator. The adder behavioral representation described above can therefore be simulated, as a stand-alone unit or along with other units in the same system, to verify its correctness. Simulation plays an important role in the design process to ensure the correctness of the intermediate and final structural (architectural) representations.

The ultimate target of a VLSI design process is to produce a set of physical masks of the circuit so that it can be fabricated at the silicon foundry. Toward this

---

[3] See the end of this chapter for some excellent books on VHDL and Verilog.

end, the behavioral representation of a circuit has to be converted to these masks by a synthesis process. Both manual and computer-aided syntheses are possible. A computer-aided synthesis process is conceptually similar to the compilation of a high level programming language into machine code and thus is occasionally referred to as silicon compilation.

We must now synthesize the architecture of the adder. The adder behavioral representation did not specify how the function s <= a + b is to be implemented in hardware. The first step in the synthesis process is thus to select an architecture (or in a more abstract sense, an algorithm) to perform the addition. A number of candidate architectures can be used to find the sum of two unsigned integer numbers.

For example, the sum can be formed with a ripple-carry addition structure, or for a better speed, with a carry-look-ahead addition structure. Still more possible addition structures exist. In either case, the structural representation of the adder is a description of the building blocks needed to implement the selected structure. According to the hierarchical design methodology, any building block can be built out of other building blocks. This relation repeats until all building blocks are manageable basic units.

In the rest of this overview, we assume that the ripple-carry addition structure illustrated in Fig. 1.5, which consists of 32 1-bit full adders (FAs), is chosen. The same information can also be conveyed with a hardware description language model (see Problem 1.6).

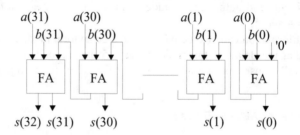

Fig. 1.5 Ripple-carry adder.

Logic Design

Moving down the building block hierarchy, we now work on the 1-bit full adder. We first present the behavioral representation of this building block.

Fig. 1.6 shows the input/output (I/O) interface of a 1-bit full adder. The full adder sums up three input bits (a, b, and c_{in}) and produces a sum bit (s) and a carry-out bit (c_{out}).

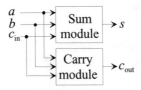

Fig. 1.6 I/O interface of a 1-bit full adder.

There are a variety of ways to describe the behavioral model of this full adder. For example, we can represent the sum function as the Boolean equation:

$$s = a \oplus b \oplus c_{in} \qquad (1.1)$$

Alternatively, we may want to represent the carry function as the following truth table:

a	b	c_{in}	c_{out}
0	0	0	0
0	0	1	0
0	1	0	0
0	1	1	1
1	0	0	0
1	0	1	1
1	1	0	1
1	1	1	1

The behavioral and structural representation of the full adder can also be described in VHDL as follows.

```
-- Interface of a full adder (Part 1)
ENTITY full_adder IS
   PORT (a, b, cin: IN BIT; s, cout: OUT BIT);
END full_adder;

-- Behavioral representation of a full adder (Part 2)
```

```
ARCHITECTURE behavioral OF full_adder IS
-- carry() and sum()
  FUNCTION carry(a, b, cin: BIT) RETURN BIT IS
    BEGIN
       RETURN (a AND b) OR (b AND cin) OR (a AND cin);
    END carry;
  FUNCTION sum(a, b, cin: BIT) RETURN BIT IS
    BEGIN
       RETURN a XOR b XOR cin;
    END sum;
-- Relationship between the outputs and inputs
-- with propagation delay information
BEGIN
  s <= sum(a, b, cin) AFTER 1 ns;
  cout <= carry(a, b, cin) AFTER 0.8 ns;
END behavioral;


-- Structural representation of a full adder (Part 3)
ARCHITECTURE structural OF full_adder IS
-- Component list
COMPONENT xor_cell
  PORT (a, b: IN BIT; z: OUT BIT);
END COMPONENT;
COMPONENT or_cell
  PORT (a, b: IN BIT; z: OUT BIT);
END COMPONENT;
COMPONENT and_cell
  PORT (a, b: IN BIT; z: OUT BIT);
END COMPONENT;

-- Sources of components
FOR ALL:xor_cell USE ENTITY WORK.lib_xor(structural);
FOR ALL:or_cell USE ENTITY WORK.lib_or(structural);
FOR ALL:and_cell USE ENTITY WORK.lib_and(structural);

-- Internal wires
SIGNAL wire1, wire2, wire3, wire4, wire5: BIT;

-- Connections
BEGIN
  a1: xor_cell PORT MAP(a, b, wire1);
  a2: xor_cell PORT MAP(wire1, cin, s);
  a3: and_cell PORT MAP(a, b, wire2);
  a4: and_cell PORT MAP(a, cin, wire3);
  a5: and_cell PORT MAP(b, cin, wire4);
  a6: or_cell PORT MAP(wire2, wire3, wire5);
  a7: or_cell PORT MAP(wire4, wire5, cout);
END structural
```

Part 1 of the above VHDL code defines the interface of a 1-bit full adder which has 3 input bits (a, b, and cin) and 2 output bits (s and cout). Part 2 is a behavioral representation showing the logical relationships between the input and output bits. Note the definitions of functions carry() and sum() as well as the use of function calls along with propagation delays.

Part 3 is a structural representation of the 1-bit full adder. It begins with a list of components (building blocks) (xor_cell, and_cell, and or_cell) and continues to specify the sources of these components (WORK.lib_xor, WORK.lib_and, and WORK.lib_or). A number of internal signal wires are defined, which is followed by a description of how the building block instances are connected. For example, the statement

```
a2: xor_cell PORT MAP (wire1, cin, s);
```

prescribes the use of a building block xor_cell with its ports a, b, and z connected to signals wire1, cin, and s, respectively. The reader is encouraged to convert the above structural representation into a logic gate schematic diagram (see Problem 1.5).

Physical Design

The physical design step is concerned with the creation of mask layouts. A set of mask layouts for a building block or even a complete IC can be manually implemented. This approach, called a full-custom design, creates a layout of geometrical entities indicating the transistor dimensions, locations, and their connections.

Computer-aided-design (CAD) tools have been developed to facilitate the design of full-custom ICs. For example, a design rule checker can be used to verify that a layout conforms to the design rules, and a routing tool can be employed to perform the wiring of the transistors.

The advantage of a full-custom design is that it allows the designer to fully control the circuit layout so that it can be optimized. However, these benefits only come at the cost of a very high design complexity. The full-custom design approach is thus usually reserved for small circuits such as the library cells to be described below, and the performance-critical part of a larger circuit. In some cases when a circuit such as a microprocessor is to be mass-produced, it may be worth the many man-months necessary to lay out a chip with a full-custom approach to achieve optimized results.

With millions of transistors involved, it is extremely difficult to manually lay out the entire chip. Another approach to converting a circuit into a physical layout uses a library of basic logic circuits, called standard-cells, in the process. This is not unlike the design of a printed circuit board system using components from standard logic families.

Standard-cell libraries are commercially available for specific fabrication processes. Sources of cell libraries are silicon foundries and independent design houses. A library usually contains basic logic gates (e.g., inverter, NAND, NOR, etc.), which can be assembled by the designer to form desired functions, and pre-designed, ready to use functional blocks (e.g., full-adder, register, etc.). According to the standard-cell approach, the behavioral representation of a design is synthesized into a structural representation that consists of only standard-cells available in the library. A layout is then created by arranging and interconnecting the standard-cells required in the design.

Compared to the full-custom approach, the complexity of designing a circuit is greatly reduced in the standard-cell approach since the designer can view the standard-cells as "black-box" components. For example, consider a cell designed in the standard-cell style, of which the physical design is shown in Fig. 1.7a.[4] The designer only needs the cell dimensions (height and width) and the locations of the signal and power lines to use it in a physical design process. The designer's view of the cell in Fig. 1.7a is shown in Fig. 1.7b. This view hides all the transistors and shows only the input/output signal lines (A, B, C, D, and F) and the power lines (V_{DD} and V_{SS}). This cell is designed to allow all signal connections to be made at both the top and bottom edges of the cell. The power lines are accessible on both sides of the cell. Please note that standard-cells also allow signal connections to be placed internal of the cell. We will further discuss these design styles in Chapter 3.

Standard-cells in a library are typically designed to have an identical height, which is called a standard height. In contrast, the widths of standard-cells are determined by their functions. A more complicated standard-cell needs a larger area so its width is longer than that of a less complicated one. The identical heights of standard-cells allow them to be conveniently arranged in a row when a physical layout is created, as shown in Fig. 1.8.

The connections to a standard-cell may also be standardized. We have shown an example of standardized connections in Fig. 1.7b. In this case, the power lines (V_{DD} and V_{SS}) of a standard-cell run horizontally on the top and the bottom of the cell, respectively. With this property, all the standard-cells in a row have their power lines abutted and connected.

If the signal wires (e.g., I/O, clock, etc.) of a standard-cell are brought from the interior of the cell to its top and/or bottom edges, the interconnections between standard-cells can also be made at predetermined locations, called pins, along the top and bottom edges of a standard-cell.[5] This property allows the interconnections between standard-cells to be made readily at their cell boundaries since the location of a signal wire can be determined by simply specifying its pin number.

---

[4] We discuss the details of creating physical designs in Chapter 3.

[5] Not to be confused with the pins on an IC package.

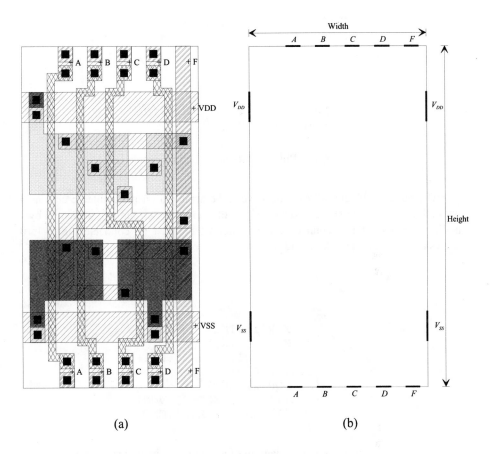

(a)   (b)

Fig. 1.7   (a) Physical design; (b) Designer's view of a standard cell.

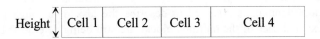

Fig. 1.8 Arrangement of standard-cells in a row.

For example, the standard-cell in Fig. 1.7b has enough width to hold twelve pin positions, six at the top and six at the bottom. Only ten of these pin positions are used. The two leftmost pin positions, one at the top and one at the bottom, are not in use. Another example of pin arrangement at the boundary of a standard-cell is shown in Fig. 1.9. Out of the eight positions available at the top of the cell, only pins 1, 4, and 8 are connected to the internal structure of the cell. Also, note that only pins 2, 4, and 7 are used at the bottom of the cell.

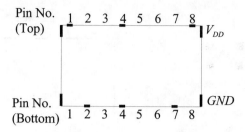

Fig. 1.9 Pin numbers in a standard-cell.

The building blocks of a circuit have to be arranged on the available chip area with all the associated signals connected. At this point, the circuit is represented by a netlist — a list of interconnections (i.e., nets) that must be connected. Fig. 1.10 shows a circuit diagram along with its netlist.

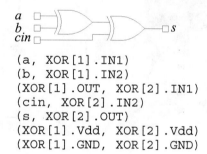

```
(a, XOR[1].IN1)
(b, XOR[1].IN2)
(XOR[1].OUT, XOR[2].IN1)
(cin, XOR[2].IN2)
(s, XOR[2].OUT)
(XOR[1].Vdd, XOR[2].Vdd)
(XOR[1].GND, XOR[2].GND)
```

Fig. 1.10 Logic diagram of a circuit and its netlist.

The task of arranging the building blocks on the layout, called the floorplanning or placement process, attempts to determine the best location for each building block. The term floorplanning is reserved to describe the task of a rough placement, in which the exact shapes of the building blocks are still being determined. The criteria for judging a placement result include the overall area of the circuit and the estimated interconnection lengths (which determine propagation delays). The routing process takes the result of a placement and automatically completes the interconnections. We describe placement and routing processes in Chapter 10.

The computational complexity of the placement and routing process goes up exponentially as the problem size (e.g., the number of building blocks) increases. The placement and routing problems, as well as many other VLSI design problems, belong to the class of combinatorial optimization problems. In a combinatorial optimization problem, the only way one could identify the truly optimal solution would be to try out all possible solutions — in general, an impossible task. Designers typically settle for pseudo-optimization by acquiring sufficiently good results within a limitation set by the available time and effort.

The standard-cell layout style allows the placement and routing of a circuit to be highly simplified. Fig. 1.11 shows an example of placement and routing of a standard-cell circuit. It has been noted above that since all cells have the same height, the placement tool readily places them in rows. All the cells in the same row have their power lines automatically connected.

The space reserved between two rows of cells, called a channel, is used for the routing purpose. There are many CAD algorithms, called channel routers, developed to perform routing in a channel. Some of them will be described in Chapter 10.

The height of a routing channel is adjustable by varying the separation distance between two rows of cells. The availability of at least two routing layers (with insulation between them) allows unrelated interconnections to cross within the channel. In the illustration of Fig. 1.11, each interconnection running in the channel consists of vertical and horizontal segments. All horizontal segments run in one metal layer and all vertical segments run in another metal layer. Vias are provided to connect them together.

If there is a need to route a signal from one channel to another, a special feedthrough cell, which provides a conducting path from its top edge to its bottom edge, can be used. The use of a feedthrough cell is demonstrated in Fig. 1.11.

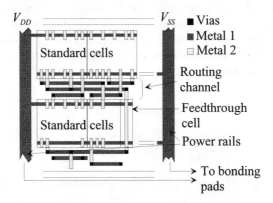

Fig. 1.11 Placement and routing of standard-cells.

Fig. 1.11 also shows the distribution of power in a standard-cell layout. Two main power rails (V_{DD} and V_{SS}) run vertically on opposite sides of the cell rows. They connect the power lines of the standard-cell rows to an external power supply.

If a fabrication technology provides more than two routing layers, a number of them can be reserved exclusively for the interconnections between cells. This allows a large amount of or all inter-cell connections to be routed above the cells to form over-the-cell routing. The pure routing area (channels and feedthroughs), which does not contribute to the circuit density, can thus be minimized.

The benefits of a standard-cell design come from the restrictions imposed on the building blocks. The standard height requirement has limited the functions available in a standard-cell library since the implementation of a complex function could require a very wide cell. The use of very wide cells is inefficient in the placement process. Another limitation of standard cells is in their performance since they were designed to meet the requirements of a range of applications.

When it is desired to use a building block with a more complex functionality, one can create a macro-cell (custom cell) that has no dimensional restriction posed on its height and width. With the requirement of having a fixed height removed, the design of a macro-cell is more flexible so a more complicated functionality is possible. Registers, register files, arithmetic-logic units, multipliers, random-access memories, read-only memories, among others, are candidates for macro-cells. While it seems natural to design macro-cells with the full-custom approach for optimal results, standard-cells can also be assembled into macro-cells.

The physical layout of a circuit using macro-cells is similar to that of standard-cells except the building blocks can no longer be efficiently placed in rows. This results in a more complicated placement and routing procedure. Fig. 1.12 shows a macro-cell based floorplan. After placement, the space between the building blocks is partitioned into rectangular regions called switchboxes. The routing of macro-cells is carried out in two steps. The first step, called global routing, determines approximately how the nets should be connected through the switchboxes. This is followed by a detailed routing step, called switchbox routing, which operates similarly to channel routing except pins are located on all four sides of a switchbox. Fig. 1.13 shows an example of switchbox routing. Similarly, if certain routing layers are reserved for inter-cell connections, many of them can be routed above the macro-cells to reduce the pure routing areas.

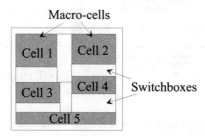

Fig. 1.12 Floorplan of a layout using macro-cells.

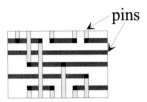

pins

Fig. 1.13 Switchbox routing example.

Bringing the macro-cell concept one step ahead, more complicated building blocks such as microprocessors, digital signal processors, memory modules, etc. can be integrated on the same chip. These building blocks are called intellectual property cores (IP cores), which can be optimized, verified, and documented to allow efficient reuses. IP cores can exist in different formats. A hard IP core is the mask information of the circuit, custom built, optimized, and verified for a specific application. A soft IP core is the behavioral description (e.g., VHDL) of a circuit, which can be parameterized and synthesized for different technologies. A hard IP core usually provides a better performance than a soft IP core but the latter is more flexible. A middle ground between hard IP core and soft IP core can be taken to produce firm IP cores. The firm IP core approach provides a register transfer level (RTL) description of a circuit, which specifies the operations to be performed on operands stored in registers. A design created using IP cores is called a system-on-a-chip (SOC) design, which has the advantage of a shorter design time by allowing previous designs to be reused.

Simulation

A chip design must be simulated many times at different levels of abstraction to ensure its correctness before it is ready for fabrication. The objective of simulation is to verify the functionality of a circuit and to predict its performance. As a circuit can be represented at different levels, circuit simulation can also be done at multiple levels ranging from transistor level to behavioral level. The transistor level simulation provides the most accurate result. However, the accuracy is only obtained at the cost of a more time consuming simulation.

For small circuits, transistor level simulation can be used and the basic elements in such a simulation are transistors, resistors, capacitors, and inductors (included only for analog circuits). This level of simulation gives the most accurate result at the cost of long simulation times.

Alternatively, transistors can be modeled as switches with propagation delays in a switch level simulation. This level significantly reduces the simulation time and frequently yields acceptable results.

If there is no need to determine the internal condition of a building block, it can be modeled at the behavioral level. Mixed level simulation, in which different parts of a system are represented differently, is commonly used to analyze a complex system. The critical parts of the system can be modeled at transistor or switch level while the rest of the circuit can be modeled at behavioral level.

Fabrication

When the layout of a circuit is complete, it is sent to a silicon foundry for fabrication. The details of silicon processing are described in Chapter 3. Fig. 1.14 illustrates a highly simplified IC fabrication process. The layout of a circuit submitted for fabrication is replicated in an array on a set of masks. Each mask is used in a photolithographic processing step to delineate a transistor feature or an interconnection feature on the wafer. This photolithographic processing step is repeated with different masks until all transistors and their connections are formed. The finished wafer is tested, sawed into individual chips, and packaged.

The fabrication process described above may be prohibitively expensive, both in terms of fabrication cost and time, for circuits which are unlikely to be manufactured in large quantities. Many VLSI designs are in the category of application-specific integrated circuits (ASICs). An ASIC is a circuit which performs a specific function in a particular application. In the ASIC market, it is important to reduce the manufacturing cost and the time to market. Special architectures have been created for these ASIC designs.

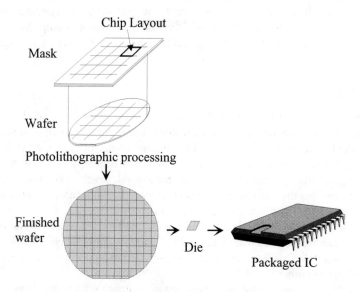

Fig. 1.14 Simplified fabrication process of an integrated circuit.

One of these architectures is the mask-programmed gate-array (or simply gate-array), which consists of an array of unconnected gate cells prefabricated on a chip. Fig. 1.15 shows the floorplan of a gate array. The ring of bonding pads at the chip boundary is provided for package connections, of which the discussion is coming up shortly. The example of a gate cell is shown in Fig. 1.16. In forming an ASIC, one can personalize a gate array by adding appropriate metal interconnections to it. For example, Fig. 1.17 shows the customization of a gate cell into a two-input NAND gate.

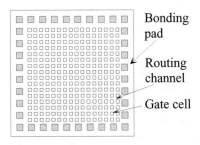

Fig. 1.15 Floorplan of an MPGA.

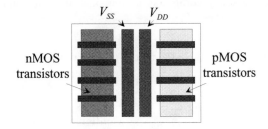

Fig. 1.16 Example of a gate cell in a gate array.

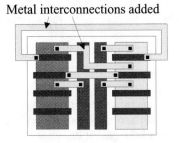

Fig. 1.17 Customization of a gate-cell.

Since all layers other than the metal layers in a gate-array are prefabricated, the turn-around time, defined as the time elapsed between the submission of a design and the receipt of chips, is reduced. The cost of a gate-array is also lower than that of a full-custom chip since gate-arrays can be mass produced and used in many different designs. Only a small number of processing steps are involved in a personalization. Another non-trivial advantage of gate-arrays is that they are often manufactured with the most advanced technology (see Problem 1.12). However, designs using gate-arrays are more restrictive than full-custom designs since all transistors are of fixed sizes.

In 1985, Xilinx introduced a gate-array structure that is programmable by the end users. To distinguish its product from those requiring foundry personalization, Xilinx called its product Field Programmable Gate-Array (FPGA). Different companies such as Actel, Altera, etc. have also developed their versions of the FPGA. Fig. 1.18 shows the floorplan of a typical FPGA. Each configurable logic block can be programmed by the user to implement any logic function of its inputs. A register is included in the configurable logic block to facilitate the implementation of sequential logic. The I/O blocks can be configured into an input terminal, an output terminal, or a bi-directional terminal. The routing between I/O blocks and logic blocks, as well as among the logic blocks themselves, is provided by the configurable interconnects which run in the routing channels between logic blocks. An FPGA is configured by downloading configuration bits that describe and program logic functions and interconnections.

Configurable I/O blocks and bonding pads

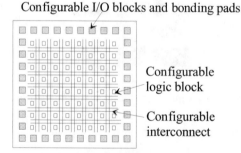

Configurable
logic block

Configurable
interconnect

Fig. 1.18 Floorplan of a typical FPGA.

Packaging

One important area in IC manufacturing is packaging. Connections (e.g., power, signals, etc.) between the circuit and the outside world must be provided. Fig. 1.19 illustrates that a chip consists of the circuit design itself and a ring of bonding pads, called a pad-frame. The signal and power connections of the circuit are connected to the bonding pads. The size of a bonding pad is typically 100×100 μm^2. This relatively large size is necessary to accommodate the operating tolerances of automatic bonding equipment. A signal buffer and a protection circuit are provided for each connection between a bonding pad and the circuit. The protection circuit protects the circuit against potential damages (e.g., static electricity).

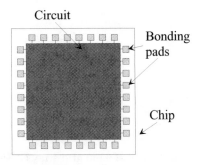

Fig. 1.19 Chip layout including a circuit and a ring of bonding pads (pad frame).

Fig. 1.20 shows the use of bonding wires to connect bonding pads to the pins of an IC package after the chip is mounted inside. In order to fully utilize the processing power of a chip, enough I/O pins must be provided. The package shown in Fig. 1.20 is called a dual-in-line package, which has a severe limitation on the number of pins available. VLSI chips require other packages that can provide more pins.

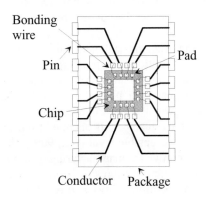

Fig. 1.20 Chip mounted in an IC package.

Fig. 1.21 shows a pin grid array (PGA) package, which arranges pins at the bottom of the package and can provide 400 or more pins. A more advanced package called a ball grid array (BGA) package has a similar array but has replaced the pins with solder bumps (balls). Another important issue with packaging is its capability of heat dissipation. Manufacturer's specifications should be carefully evaluated before a package is selected.

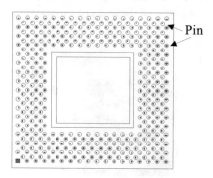

Fig. 1.21 Bottom view of a pin grid array (PGA) package.

Testing

Manufactured chips normally have to go through at least two tests before acceptance. There are two principal causes of failure, design errors and manufacturing defects. A wafer test is applied to visually check, identify, and mark faulty chips. The wafer is then sawed into individual circuit chips and the marked faulty chips are discarded. Each good chip is mounted on a supporting base of an IC package. Wires are used to connect the bonding pads on the chip and the pins on the package.

The packaged ICs then undergo a functional test to separate acceptable parts from failed ones. In the fabrication of a prototype, analysis must be performed to determine the causes of detected failures. After design errors have been identified and eliminated, the chip goes into production. During IC production, a certain number of chips can be expected to fail the acceptance test and be discarded along the way. The percentage of acceptable parts obtained from a wafer is called the yield. If the yield drops below an acceptable level (e.g., 90%), the wafer processing steps should be investigated to improve the manufacturing quality.

The functional test of a circuit requires its internal nodes to be accessible so that test signals can be injected (controllability), and responses at the internal nodes can be monitored (observability). Unlike a printed circuit board, an integrated circuit has a very limited degree of controllability and observability. Accordingly, VLSI design methodology must embody the important concept of design-for-testability (DFT). Instead of considering the testing of a VLSI circuit as a step isolated from the rest of the design process, the need for testing has to be taken into account throughout the design process. Special approaches have been developed to improve the testability of circuits. For example, registers can be placed at critical testing points so that they can be cascaded into a scanning shift register to permit the control and observation of internal signals. An approach called boundary scan, which builds circuitry into an integrated circuit to assist in the test, maintenance, and support of assembled printed circuit boards, has been developed into IEEE standard 1149.1.[6]

Another problem facing designers is the complexity of testing. Consider the testing of a combinational logic circuit. If this circuit has n inputs, there are 2^n possible input combinations. A straightforward testing approach is to apply each of these input combinations (test vectors) in sequence to check out the circuit. This testing approach, called exhaustive testing, is only possible for very small circuits. The problem is further complicated if a sequential circuit is to be tested. The outputs of a sequential circuit are functions of not only its current inputs, but also its history of operations. The objective of a test scheme development for a circuit is to identify the smallest set of test vectors necessary to verify the circuit functions. Testing of VLSI circuits, which has been a major research topic in the field of VLSI engineering, will be discussed in Chapter 9.

1.3   Summary

In this chapter the typical design steps for producing a VLSI circuit have been presented. Fig. 1.22 summarizes these steps in a flow chart. Two design flows are evident in the chart. The top-down design flow takes an abstract, behavioral representation of the design target and synthesizes it into a structural representation which consists of a set of building blocks and their interconnections. A behavioral representation generally has a number of different, yet functionally equivalent implementations. Mixed level simulation can be used to verify and evaluate different implementations.

The second half of the design process, which is a bottom-up design flow, takes the structural representation and transforms it into a physical layout. A full-custom design uses hand-crafted layouts to maximize the performance, but its complexity restricts its use to small circuits and to performance critical portions of a larger system. A standard-cell design takes advantage of a pre-designed, universal standard-cell library to reduce design complexity. A number of features, such as the identical heights of the cells in a library and the standardized I/O and power rail locations, allow efficient placement and routing tools to be developed.

---

[6] IEEE: Institute of Electrical and Electronics Engineers (http://www.ieee.org).

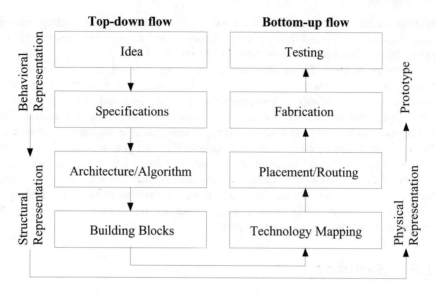

Fig. 1.22 VLSI circuit design flow.

Instead of fabricating a circuit from scratch, the mask-programmed gate array provides quick turn around time by prefabricating an array of unconnected transistors on a wafer. It is then personalized by adding the metal interconnections according to the specific implementation. The field programmable gate-array (FPGA) goes even further by providing a fully packaged device that is user programmable for application specific integrated circuits (ASICs).

Testing has a very important role in the VLSI design process. Due to the limited controllability and observability of VLSI circuits, testability has to be emphasized throughout the design process instead of being treated as an afterthought.

1.4   To Probe Further

An overview of many VLSI design issues has been introduced in this chapter. While many of these subjects will be discussed further in later chapters, it is impossible for a book of this size to give an in-depth coverage to every one of them. Furthermore, the VLSI design methodology and IC fabrication technology advance rapidly. We have thus provided at the end of each chapter a list of resources to probe further on the subjects discussed in that chapter.[7]

---

[7] Reasonable effort has been used to verify the web sites listed in this book. However, please note that the availability of a web site and its contents are beyond our control. The inclusion of a commercial web site does not represent any form of endorsement from the author or the publisher. All product names mentioned are the trademarks of their respective owners.

International Technology Roadmap for Semiconductors:

- A lot of information about developments in semiconductor technologies can be found in the web site: http://www.itrs.net/ntrs/publntrs.nsf

Moore's Law:

- R. R. Schaller, "Moore's Law: past, present, and future," *IEEE Spectrum*, June 1997, pp. 53-59.

VHDL:

- J. Bhasker, *A VHDL Primer, 3rd Edition.*, Prentice-Hall, 1999.

- J. R. Armstrong, *Chip-Level Modeling With VHDL*, Prentice-Hall, 1989.

Verilog:

- D. E. Thomas, and P. Moorby, *The Verilog Hardware Description Language*, Kluwer Academic Publishers, 1991.

FPGAs:

- S. M. Trimberger, ed., *Field-Programmable Gate Array Technology*, Kluwer Academic Publishers, 1994.

- http://www.xilinx.com

- http://www.altera.com

- http://www.actel.com

Microprocessors/Digital Signal Processors:

- http://www.intel.com

- http://www.motorola.com

- http://www.ti.com

Silicon Foundries:

- http://www.mosis.org

- http://www.tsmc.com

- http://www.umc.com

IP Cores/SOC:

- http://www.design-reuse.com

- H. Chang et al., *Surviving the SOC Revolution: A Guide to Platform-Based Design*, Kluwer Academic Publishers, 1999.

- L. M. Silveira, S. Devadas, and R. Reis, Ed., *VLSI: Systems on a Chip*, IFIP TC10 WG10.5 Tenth International Conference on Very Large Scale Integration (VLSI'99), December 1-4, 1999, Lisboa, Portugal, Kluwer Academic Publishers, 1999

CAD Tools:

- http://www.mentor.com

- http://www.cadence.com

- http://www.synopsys.com

- http://www.avanticorp.com

Journals, Magazines, and Proceedings:

- http://www.ieee.org

- http://www.acm.org

- ACM/IEEE Design Automation Conference, Proceedings (http://www.dac.com)

- Advanced Research in VLSI, Proceedings

- Great Lakes Symposium on VLSI, Proceedings

- IEEE Circuits and Devices Magazine

- IEEE Custom Integrated Circuits Conference, Proceedings

- IEEE Design & Test of Computers

- IEEE International Conference on Computer-Aided Design, Digest of Technical Papers (www.iccad.com)

- IEEE International Symposium on Circuits and Systems, Proceedings

- IEEE Journal of Solid-State Circuits

- IEEE Spectrum

- IEEE Transactions on Circuits and Systems

- IEEE Transactions on Computer-Aided Design of Integrated Circuits and Systems

- IEEE Transactions on Very Large Scale Integration (VLSI) Systems

- International Conference on Computer Design, Proceedings

- International Test Conference, Proceedings

1.5   Problems

1.1   According to Moore's law, what would be the potential capacity of a memory chip in 2005 if the most advanced technology in 2000 produces 1 giga-bits per memory chip?

1.2   Many researchers currently predict that "the lower bound of feature size is around 0.05 μm for the current form of MOSFET ICs." Research and find out the problems a fabrication process would have to cope with at that feature size.

1.3   Research and explain why BJT-based logic circuits have higher speed and larger current capability than MOSFET-based counterparts.

1.4   Based on the information provided in this chapter and your personal experience, describe the similarities and differences between designing an integrated circuit and a printed circuit board.

1.5   Draw a gate level schematic diagram of the 1-bit full adder implementation described by the VHDL structural representation listed in page 10.

1.6   Write a pseudo VHDL structural representation for a 4-bit ripple-carry adder.

1.7   The Boolean equations of a 1-bit full adder are given below:
$S = A\overline{B}\,\overline{C_i} + \overline{A}BC_i + \overline{A}B\overline{C_i} + ABC_i$; $C_o = AB + BC_i + AC_i$.

Design a 1-bit full adder using each of the following approaches:
- An implementation using only 2-input NAND gates.
- An implementation using only 2-input NOR-gates.
- A read only memory (ROM) based implementation.
- A multiplexer based implementation.

Give schematic diagrams of these designs at appropriate levels. Which one would be your choice of implementation for the 32-bit ripple carry adder described in this chapter? Justify your selection by considering the following evaluation criteria:
- Design complexity.

- Cost of fabrication (PC board, chips, sockets, etc.).
- Circuit performance (power, speed, size, reliability, etc.).

1.8   You have just started to work as a VLSI designer. Your first assignment is to implement a multiplication unit. Develop specifications for this multiplication unit.

1.9   Create a flow chart to illustrate the steps involved in the design process of the multiplication unit in Problem 1.8. Show any iterations (loops) in your flow chart explicitly.

1.10   A microprocessor has a 32-bit program counter and sixteen 32-bit general registers. What is the minimum number of different states this microprocessor has? Using a tester that is capable of making 200 million tests a second, how long will it take to check out all these states?

1.11   This problem refers to the "finished wafer" shown in Fig. 1.14. Assuming all complete chips are good circuits, what is the yield of this process? Notice that your result is the upper bound of yield.

1.12   Give a reason why gate-array and FPGA designers often have access to advanced fabrication technologies earlier than full-custom designers.

1.13   Determine the number of different ways to place 5 building blocks (e.g., logic gates) of identical shapes and sizes into 5 slots (each slot can hold at most one building block) on a chip. Repeat for a chip of 10 building blocks and 10 slots. Plot the number of possible placement arrangements as a function of n, the number of building blocks.

Chapter 2 CMOS Logic Circuits

Logic circuits with a touch of CMOS ...

CMOS logic circuits are made of n-channel and p-channel MOSFETs. Unlike their bipolar transistor based counterparts (e.g., emitter-coupled logic circuits), CMOS logic circuits can be built exclusively with transistors and do not require the use of resistors and diodes. A MOSFET has a very high resistance when it is turned off. This unique property allows the designer to use highly simplified MOSFET switch models in the design and analysis of CMOS logic circuits.

Sophisticated MOSFET models have been developed for and used in computer-aided circuit simulators to accurately simulate MOSFET circuits. We will see in Chapter 4 that such elaborate MOSFET models are indispensable in the determination of logic circuit transient behaviors. The "drawback" of these models, however, is that their usage involves a large amount of complicated computation.

For example, the MOSFET model employed in the popular circuit simulator SPICE has more than 40 parameters.[1] We introduce a SPICE model for a MOSFET in Chapter 4. The readers who have experience with SPICE can testify that the analysis of a circuit consisting of just a few hundred transistors can be a very time consuming process. A simplified and efficient modeling approach is definitely needed to tackle VLSI design problems involving the simulation of millions of transistors.

Initially, we model a MOSFET as a simple digital-signal-controlled switch. We then demonstrate the use of this simple switch model in the design and analysis of CMOS logic functions. This switch model will be enhanced in Chapter 4 by incorporating with it some non-idealistic MOSFET behaviors. The enhanced model can be used to estimate the propagation delay and power dissipation of CMOS circuits.

2.1   nMOS Switch Model

Fig. 2.1 shows the physical structure of an n-channel MOSFET, hereinafter referred to as an nMOS transistor, along with its circuit symbol. The nMOS transistor is fabricated on a p-type substrate. Two heavily doped n^+ regions are created in the substrate to form a source (S) and a drain (D). The structure of a MOSFET is symmetrical so the source and drain of an nMOS transistor are interchangeable with no changes in device characteristics. However, for the sake of circuit analysis, one of

---

[1] SPICE (Simulation Program with Integrated Circuit Emphasis) was originally developed by the Integrated Circuits Group of the Electronics Research Laboratory and the Department of Electrical Engineering and Computer Sciences at the University of California, Berkeley, California. SPICE has since become the synonym of computer-aided circuit analysis.

the n$^+$ regions, usually the one at a potential higher than another, is labeled as the drain.

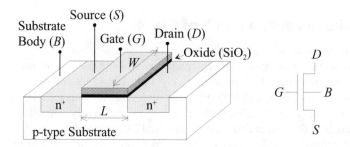

Fig. 2.1 Physical structure and circuit symbol of an nMOS transistor.

As shown in Fig. 2.1, a polysilicon gate (G) sits on top of a channel area (the area between the two n$^+$ regions) with a thin (on the order of 10^{-2} µm) layer of insulating silicon dioxide (SiO$_2$) sandwiched in between.[2] Historically, the gate was made of metal. The name metal-oxide-semiconductor (MOS) reflects the structure consisting of metal (gate), oxide (silicon dioxide), and semiconductor (silicon). Currently the gate is no longer made of metal but the name MOSFET has survived.

From the viewpoint of VLSI designers, an nMOS transistor is characterized by its channel length (L) and width (W) defined by the dimensions of its gate (see Fig. 2.1). Since the switching speed of a transistor is inversely proportional to its channel length, it is usually desirable to keep the channel length L as short as possible. In most cases, the channel length L is set at the minimum feature size of the fabrication process — the width of the narrowest wire that the process can produce. The most advanced fabrication process in 2000 can produce a feature size of about 0.15 µm.

The relationship between the channel width and the transistor characteristic is more complicated. We will get back to this subject in Chapter 4. It is sufficient to understand for now that the channel width W is a parameter selectable by the designer to produce desired circuit properties. The channel width ranges from below one micron to several hundred microns.

Fig. 2.1 also shows that an nMOS transistor has a fourth terminal that is connected to the substrate body (B). The substrate must be connected to the lowest potential (V_{SS} or GND) in a circuit to prevent the p-n junctions (i.e., diodes) formed between the n$^+$ regions and the p-type substrate from being forward-biased and turned on. A forward-biased p-n junction in a MOSFET, even if it happens momentarily, may induce a serious problem called latch-up. We will further discuss the latch-up problem in Chapter 4.

All nMOS transistor substrate body contacts in an IC are connected together and set at an appropriate potential. This arrangement allows the simplified circuit symbol

---

[2] Polysilicon is non-crystalline silicon doped to act as a conducting material.

in Fig. 2.2 to be used. The symbol depicts an nMOS transistor as a three terminal device and omits its body terminal.

Fig. 2.2 Simplified nMOS transistor symbol.

In a highly simplified view, an nMOS transistor operates in one of two states. If the voltage between its gate and source (v_{GS}) is below a positive threshold voltage (V_{tn}), the transistor is in a cut-off state so that a high resistance (on the order of 10^{12} Ω) exists between its drain and source. The threshold voltage V_{tn} is a physical property of the transistor and is usually between 0.5 V and 1 V. Alternatively, if $v_{GS} > V_{tn}$, the nMOS transistor turns on to provide a conducting path (on the order of 10^3 Ω) between its drain and source.

The nMOS switch model can be derived by considering the nMOS circuit shown in Fig. 2.3. Only $v_I = V_{DD}$ (the power supply node, a "high" voltage above V_{tn} representing logic-1) and $v_I = V_{SS}$ (the ground, a "low" voltage below V_{tn} representing logic-0) are considered in this nMOS switch model. The power supply V_{DD} ranges from 5 V to 1 V. The use of a lower V_{DD} in a CMOS circuit decreases its power dissipation at the cost of a lower speed as well as a reduced noise margin. Experimental low power systems using a V_{DD} as low as 400 mV have been reported.

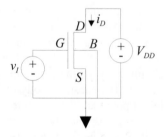

Fig. 2.3 Circuit for deriving the nMOS switch model.

V_{SS} is also called *GND* and is typically set at 0 V. When a zero bias voltage ($v_I = 0$ or logic-0) is applied to the gate, $v_{GS} = 0 < V_{tn}$ so the nMOS transistor is in its cut-off state ($i_D \cong 0$). Alternatively, if v_I is raised to V_{DD} (logic-1) so that $v_{GS} > V_{tn}$, the nMOS transistor turns on ($i_D > 0$). Eventually we will improve this switch model by incorporating an effective channel resistance, but at this point we simply assume that the transistor is a perfect switch (i.e., no resistance) when it is turned on. Fig. 2.4 summarizes the operation of an nMOS switch when its gate is set to logic-0 and logic-1, respectively.

$$
\begin{array}{cc}
0 & 1 \\
\downarrow & \downarrow \\
D \; \underset{\bullet\;\bullet}{\searrow} \; S & D \; \underset{\bullet\;\bullet}{\searrow} \; S
\end{array}
$$

Fig. 2.4 Two operating modes of the digitally controlled nMOS switch model.

2.2   pMOS Switch Model

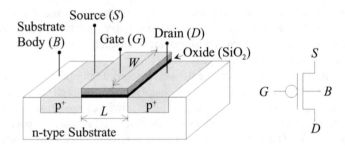

Fig. 2.5 Physical structure and circuit symbol of a pMOS transistor.

Fig. 2.5 shows the physical structure and circuit symbol of a pMOS transistor. A pMOS transistor is similar to an nMOS transistor except in the following ways. Two heavily doped p^+ regions in the n-type substrate form a source and a drain. The MOSFET structural symmetry allows the source and drain of a pMOS transistor to be interchangeable. In a circuit analysis, the p^+ region with a potential higher than another is considered the source. The substrate body of a pMOS transistor is always connected to V_{DD} or the highest potential in a circuit to reverse-bias the p-n junctions (i.e., diodes) between the p^+ regions and the n substrate. This allows the three terminal device symbol in Fig. 2.6 to be used to represent a pMOS transistor, since the substrate body is always connected appropriately.

Fig. 2.6 Simplified pMOS transistor symbol.

The most significant difference between a pMOS transistor and an nMOS transistor is that the pMOS transistor has a negative threshold voltage (V_{tp}). The typical value of the threshold voltage V_{tp} is between -0.5 V and -1.2 V. A pMOS transistor turns on when its gate to source voltage v_{GS} is below its threshold voltage V_{tp}.

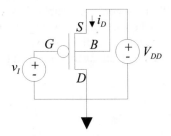

Fig. 2.7 Circuit for deriving the pMOS switch model.

A pMOS circuit is set up in Fig. 2.7 to derive the pMOS switch model. The reader should compare Fig. 2.3 and Fig. 2.7 to observe their differences in terminal label assignments and in the directions of current i_D. Again, only $v_I = V_{DD}$ (logic-1) and $v_I = V_{SS}$ (logic-0) are considered in the pMOS switch model. A voltage $v_I = 0V$ (logic-0), which is equivalent to $v_{GS} = -V_{DD} < V_{tp}$, turns on the pMOS transistor ($i_D > 0$). On the other hand, if $v_I = V_{DD}$ (logic-1), $v_{GS} = 0V > V_{tp}$ so the pMOS transistor turns off. The operation of the simplified pMOS switch model is summarized in Fig. 2.8.

Fig. 2.8 Two operating modes of the digitally controlled pMOS switch model.

In summary, the MOSFET switch model considers both the nMOS and pMOS transistors as "perfect" switches controlled by digital signals applied to their gates.

The operations of both nMOS and pMOS transistors according to the switch model are listed in Fig. 2.9.

v_I	Logic Level	nMOS	pMOS
V_{DD}	1	ON	OFF
V_{SS}	0	OFF	ON

Fig. 2.9 Switch model of MOSFETs

We can see, according to the MOSFET switch model, that nMOS and pMOS transistors are complementary switches because a logic signal that turns on an nMOS transistor turns off a pMOS transistor and vice versa. This complementary relationship is captured in the pMOS circuit symbol that includes a circle ("bubble") at the gate of an "nMOS transistor." A "bubble" is often used to represent an inverting function in digital schematic diagrams. Fig. 2.10 illustrates this relationship by showing that a pMOS switch is, in the context of switching logic, functionally equivalent to an nMOS switch with an inverter attached to its gate.[3]

Fig. 2.10 Complementary relationship between nMOS and pMOS transistors.

2.3 CMOS Inverter

In Section 2.5, we describe the design of general CMOS complementary logic circuits. In order to prepare for that, we use the idealized MOSFET switch model shown in Fig. 2.9 to design and analyze a CMOS inverter. The techniques explored in this design example can be readily extended and applied to any CMOS complementary logic circuits. In fact, we will show in Chapter 4 that all complementary logic circuits can essentially be viewed as generalized inverters.

The benefits and drawbacks of the idealized MOSFET switch model will become obvious after we study the design example of an inverter. The simplified model cannot be used to evaluate circuit performance such as delay time and power

---

[3] Of course, it is physically impossible to build a pMOS transistor this way.

dissipation. However, it does allow a desired logic function to be readily implemented.

The truth table of an inverter is given along with its logic symbol in Fig. 2.11.

A	Z
0	1
1	0

$A \; \longrightarrow \!\!\!\!\!\triangleright\!\!\!\circ\!\!- Z$

Fig. 2.11 Truth table and logic symbol of an inverter.

Throughout the book we assume positive logic signals are used; so high and low voltages are always associated with logic-1 and logic-0, respectively. We will further assume that the high voltage is V_{DD} and the low voltage is V_{SS}. In all schematic diagrams of this book we use arrows pointing upward ($\uparrow$) and downward ($\downarrow$) to graphically represent V_{DD} and V_{SS}, respectively.

Consider the first row in the inverter truth table. It states that output Z should be a logic-1 whenever input A is a logic-0. For a switch based implementation, a logic-0 input should cause output Z to connect to logic-1, or V_{DD}.[4] This operation calls for a path that conducts and connects output Z to V_{DD} when $A = 0$. This path, called a "pull-up" path since it "pulls" the output node Z "up" to V_{DD}, is illustrated in Fig. 2.12.

In Fig. 2.12a, the switch is closed when $A = 0$ so Z is connected to V_{DD} (1). In Fig. 2.12b, when $A = 1$, the switch opens. This puts output node Z in a floating (i.e., open-circuit) state, a generally undesirable state in a logic circuit. The behavior of a floating node in a CMOS circuit will be discussed in more detail.

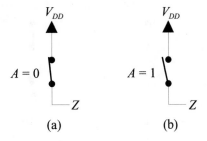

Fig. 2.12 Pull-up path of an inverter: (a) $A = 0$; (b) $A = 1$.

---

[4] From this point on, if a signal is a logic-1 or logic-0, we simply state that it equals 1 or 0, respectively.

Now consider the second row in the inverter truth table. Whenever input A equals 1, output Z should be 0. In other words, when input $A = 1$, output Z is supposed to be connected to V_{SS} (0). We call this a "pull-down" path since it "pulls" the output node Z "down" to V_{SS}. This pull-down path and its two operating states are illustrated in Fig. 2.13. Output Z equals 0 when $A = 1$, and it is in a floating state when $A = 0$.

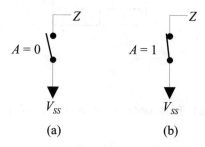

Fig. 2.13 Pull-down path of an inverter: (a) $A = 0$; (b) $A = 1$.

Finally, we put the pull-up and pull-down paths together and create the inverter switch network shown in Fig. 2.14. Replacing the ideal switches with transistors, we have the circuit shown in Fig. 2.15. Note that a pMOS transistor is used to implement the pull-up path since it turns on when $A = 0$. In the pull-down path, an nMOS transistor is used since it turns on when $A = 1$.

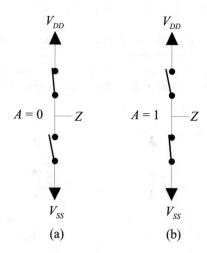

Fig. 2.14 Switch network implementing an inverter: (a) $A = 0$; (b) $A = 1$.

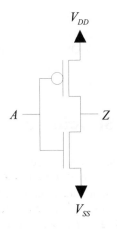

Fig. 2.15 CMOS inverter.

A logic function being designed often has several functionally equivalent implementations. The designer has to evaluate them so that an appropriate choice can be made according to the design requirements. At this time, we have yet to develop adequate models for evaluating circuit performance, but the following observations can be made. Three tangible performance measures are typically used to evaluate VLSI designs: area, propagation delay, and power consumption. Other less tangible evaluation criteria are design complexity, reliability, testability, and cost.

We discuss the physical layout and fabrication process of CMOS circuits in Chapter 3. Before a layout is produced, the number of transistors in a circuit is often used as a measure to estimate the area of a circuit. Note that while the circuit area is generally proportional to the number of transistors, it also depends on other factors such as the transistor dimensions and the interconnections between the transistors. The exact area of a CMOS circuit can only be determined after its layout has been created. The CMOS inverter in Fig. 2.15 includes two transistors.

As noted, the switch model that we used to design the inverter assumes that the transistors are perfect switches that cause no propagation delay. This assumption is apparently unrealistic; otherwise, we would have infinitely fast circuits. Our enhanced switch model, to be described in Chapter 4, will provide a simple method to estimate the speed of a circuit. Until then, we will use the number of gate levels as a measure to estimate the delay time. The CMOS inverter is a single gate and thus it has one level of delay. Naturally, different gates have different delay times so the value of this delay estimation is limited.

The switch network in Fig. 2.14 shows that the inverter does not have a conducting path between V_{DD} and V_{SS} when input A equals either 1 or 0. No conducting path means no current flow, which implies that the inverter has no power dissipation. This statement is only partially true since Fig. 2.14 does not depict the inverter transient operation — when the voltage at node A changes from 0 to 1, or vice versa. During such a transition, both transistors conduct at the same time so power dissipation occurs. We will have more discussion on this in Chapter 4. Nevertheless, we have just witnessed an important feature of the CMOS complementary logic circuit — a CMOS logic circuit does not consume power when it is in a steady state. This is one of the main reasons that the CMOS technology has been so popular for low power applications.

2.4   CMOS Logic Structures

Many CMOS structures have been used to implement logic functions. The diagram in Fig. 2.16 presents a listing of CMOS logic circuits that this book covers. In this book, we discuss two main categories of CMOS logic circuits, static logic circuits and dynamic logic circuits. A dynamic logic circuit produces its output by storing charge in a capacitor. The output thus decays with time unless it is refreshed periodically. In contrast, a static logic circuit holds its output indefinitely. In this chapter, we introduce the design of CMOS complementary logic and pass-transistor/transmission-gate logic circuits. The rest of the logic structures will be described in detail in Chapter 6.

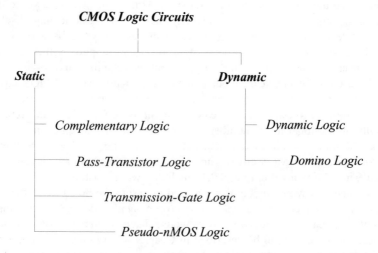

Fig. 2.16 Different structures of CMOS logic circuits.

2.5   Complementary Logic

Fig. 2.17 shows the general structure of a CMOS complementary logic circuit. It contains two complementary transistor networks. The complementary pull-up and pull-down transistor networks, controlled by the same set of inputs, selectively connect output Z to V_{DD} and V_{SS}, respectively. Due to the complementary relationship between the pull-up and pull-down networks, an input combination that turns on the pull-down network turns off the pull-up network, and vice versa. This eliminates the possibility of having a floating output Z. We can see that the inverter is indeed an example of a CMOS complementary logic circuit in which each transistor network consists of only a single transistor.

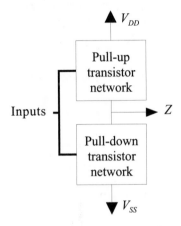

Fig. 2.17 General structure of CMOS complementary logic circuit.

Fig. 2.18 Serial switch network.

Switches can be combined in various ways to produce desired logic operations. For example, consider the serial switch network shown in Fig. 2.18. We observe that the network conducts only if switches A and B are both closed. Assume that switches A and B are implemented with nMOS transistors which are turned on by $A = 1$ and $B = 1$, respectively; then output Z can be written as

$$Z = X(AB) + HiZ(\overline{AB})\tag{2.1}$$

where *HiZ* (high impedance) represents the floating condition at Z when the switch network is turned off so Z is left unconnected. Alternatively, if pMOS transistors are used to implement switches A and B, which are turned on by $A = 0$ and $B = 0$, respectively, output Z takes the form of

$$Z = X(\overline{A}\ \overline{B}) + HiZ(\overline{\overline{A}\ \overline{B}}).\tag{2.2}$$

On the other hand, the parallel switch network shown in Fig. 2.19 conducts if either switch C or switch D is closed. Implementing the switches with nMOS transistors, output F can be written as

$$F = Y(C + D) + HiZ(\overline{C+D}).\tag{2.3}$$

Finally, if pMOS transistors are used in the parallel network,

$$F = Y(\overline{C}+\overline{D}) + HiZ(\overline{\overline{C}+\overline{D}}).\tag{2.4}$$

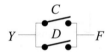

Fig. 2.19 Parallel switch network.

Based on (2.1) and (2.2), a Boolean product (AND) of variables can be formed by the serial connection of switches controlled by these variables. Also, a Boolean sum (OR) of variables can be created by the parallel connection of switches controlled by these variables according to (2.3) and (2.4). Combinations of serial and parallel connections are used to implement the pull-up and pull-down transistor networks of a CMOS complementary logic circuit.

We begin the discussion of complementary logic circuit design by stating the following guidelines:

- All switches in a pull-up network should be implemented by pMOS transistors. All switches in a pull-down network should be implemented by nMOS transistors.
- The number of switches connected in series should be kept less than 5.

The first guideline assures that the output of a logic circuit will have a maximum logic swing (V_{DD} to V_{SS}) since pMOS and nMOS transistors are good switches for

V_{DD} and V_{SS}, respectively. Reversal of their roles would produce "weak" outputs.[5] For example, using an nMOS transistor to connect an output to V_{DD} results in an output voltage that is lower than V_{DD}, a weak-1.

Fig. 2.20 illustrates a weak-1 output situation resulting from the use of an nMOS transistor. Recall that an nMOS transistor conducts only when its gate-to-source voltage v_{GS} is higher than a threshold voltage V_{tn}. The output voltage v_{CL} in Fig. 2.20 thus cannot reach beyond ($V_{DD} - V_{tn}$) since that would cause the transistor to turn off. Note that we have used a capacitor C_L to represent the output of the circuit. We will see that this is typical in a CMOS circuit.

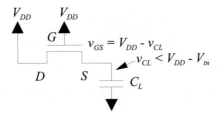

Fig. 2.20 Output connected to V_{DD} through an nMOS transistor.

Similarly, a pMOS transistor used to connect an output to V_{SS} results in a weak-0 output that cannot go below ($V_{SS} - V_{tp}$). This is illustrated in Fig. 2.21.

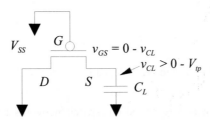

Fig. 2.21 Output connected to V_{SS} through a pMOS transistor.

Fig. 2.22 graphically illustrates the reduction of logic swing in a logic circuit due to the weak signals. In addition, weak signals increase power consumption since they may not be able to turn transistors off completely. Note that there are exceptions to this guideline. Despite the disadvantages of weak signals, a design can often be simplified by accommodating weak signals. Buffers can be used to restore signal strength as needed.

---

[5] A weak 1 signal has a voltage level that is less than V_{DD} and a weak 0 signal has a voltage level that is higher than V_{SS}. The logic swing is thus reduced.

The second guideline also arises from the desire of achieving a better performance. A signal is delayed when it passes through a transistor. The delays of serially connected transistors add up and contribute to the delay of the logic circuit.

We present a few examples to demonstrate the design of CMOS complementary logic circuits.

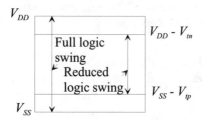

Fig. 2.22 Full logic swing and reduced logic swing.

Example 2.1

Design a CMOS complementary logic circuit to implement a two-input NAND gate $(F = \overline{AB})$, of which the logic symbol is shown in Fig. 2.23.

$$A \quad B \quad F$$

Fig. 2.23 Logic symbol of a two-input NAND gate.

As we have shown above, MOSFET switch networks are used to implement Boolean products and sums of variables. The pull-up network expression for function F is the expression F itself. We thus rewrite logic function F into a pull-up network expression that consists of only the sums and/or products of variables A, B, $\overline{A}$, and $\overline{B}$.

The structure of the pull-up network is determined by

$$F = \overline{AB} = \overline{A} + \overline{B} \tag{2.5}$$

which is a Boolean sum of $\overline{A}$ and $\overline{B}$. According to (2.4), we use a parallel network of two pMOS transistors to implement the pull-up network. The result is shown in Fig. 2.24. This network connects output F to V_{DD} (1) when $A = 0$ or $B = 0$.

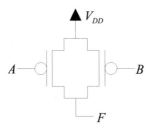

Fig. 2.24 Pull-up network for $F = \overline{AB}$.

When the pull-up network is turned off (by $A = 1$ and $B = 1$), a complementary pull-down network connected to F has to be turned on to avoid the condition of $F = HiZ$. The pull-down network for function F is determined by its complementary function $\overline{F}$. The structure of this pull-down transistor network is found to be

$$\overline{F} = AB \tag{2.6}$$

which is a Boolean product of A and B. According to (2.1), a serial switch network of two nMOS transistors is used to implement the pull-down network. This is shown in Fig. 2.25. The network turns on only if $A = 1$ and $B = 1$. When this happens, output F is connected, through the conducting pull-down network, to V_{SS} (0). We put the pull-up and pull-down networks together into the two-input NAND gate shown in Fig. 2.26.

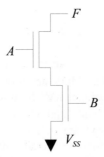

Fig. 2.25 Pull-down network for $F = \overline{AB}$.

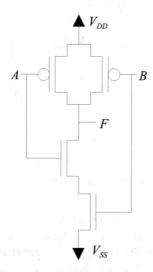

Fig. 2.26 Transistor schematic diagram of a two-input NAND gate.

Due to the complementary relationship between the pull-up network and pull-down network in a logic circuit, De Morgan's Theorem can be applied to one of the networks to derive the structure of another network. According to De Morgan's Theorem, given two variables A and B,

$$\overline{AB} = \overline{A} + \overline{B} \tag{2.7}$$

and

$$\overline{A + B} = \overline{A}\ \overline{B} \tag{2.8}$$

Based on (2.7), the complementary structure of a serial nMOS structure (AB) in the pull-down network is a parallel pMOS structure ($\overline{A} + \overline{B}$) in the pull-up network. In addition, (2.8) states that a parallel nMOS structure $(A + B)$ in the pull-down network should be matched with a serial pMOS structure ($\overline{AB}$) in the pull-up network. Similarly, it can be shown that parallel and serial pMOS structures in a pull-up network are matched with serial and parallel nMOS structures in a pull-down network, respectively. Fig. 2.27 summarizes the principles of creating a complementary transistor network for a given one. These principles are demonstrated with the design in Example 1.2.

Pull-up network	Pull-down network
pMOS transistor	nMOS transistor
serial connection	parallel connection
parallel connection	serial connection

Fig. 2.27 Complementary pull-up and pull-down networks in a logic circuit.

Example 2.2

Design a CMOS complementary logic circuit to implement a two-input NOR gate ($F = \overline{A + B}$), of which the logic symbol is shown in Fig. 2.28.

Fig. 2.28 Logic symbol of a two-input NOR gate.

The expression $F = \overline{A + B}$ is first rewritten into a product of $\overline{A}$ and $\overline{B}$ ($F = \overline{A}\overline{B}$). The pMOS serial network shown in Fig. 2.29 is used to implement this product into a pull-up network. We can now create the pull-down network by applying the principles in Fig. 2.27. According to Fig. 2.27, the two serial pMOS transistors in the pull-up network are matched with two parallel nMOS transistors in the pull-down network. The complete circuit of a 2-input NOR gate is shown in Fig. 2.30.

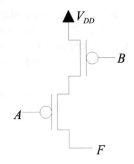

Fig. 2.29 Pull-up network for $F = \overline{A + B}$..

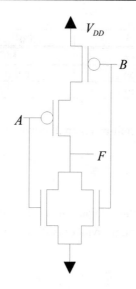

Fig. 2.30 CMOS two-input NOR gate.

The reader may have observed that, in the above example, $\overline{F} = A + B$ (i.e., the pull-down network) is naturally expressed as a Boolean sum of A and B. It is thus simpler to construct the pull-down network first and derive the complementary pull-up network using Fig. 2.27. This observation is indeed valid. In general, out of the two switching networks in a logic circuit, one would select and construct the one that is more convenient to implement. A complementary network is then derived to complete the design.

In many applications, a design problem is described in the form of a truth table or a Boolean equation. In addition to apply Boolean algebra to derive a complementary logic circuit, the Karnaugh map proves to be a valuable tool for paper-and-pencil designs. Note that the Karnaugh map method is limited to small designs with six or less variables since it becomes extremely unwieldy when more than six variables are involved. Computerized methods such as the Quine-McCluskey algorithm should be used in those cases (see references listed at the end of this chapter).

We demonstrate the use of a Karnaugh map in the derivation of a complementary logic circuit in the following example.

Example 2.3

Design a complementary logic circuit for $F(ABCD) = \Sigma m(0, 1, 2, 3, 4, 5, 6, 7, 8, 9, 10)$. Note that this expression indicates the minterms of the logic function F.

The truth table of F is given in Fig. 2.31.

A	B	C	D	F
0	0	0	0	1
0	0	0	1	1
0	0	1	0	1
0	0	1	1	1
0	1	0	0	1
0	1	0	1	1
0	1	1	0	1
0	1	1	1	1
1	0	0	0	1
1	0	0	1	1
1	0	1	0	1
1	0	1	1	0
1	1	0	0	0
1	1	0	1	0
1	1	1	0	0
1	1	1	1	0

Fig. 2.31 Truth table of F in Example 2.3.

A four-variable Karnaugh map is created in Fig. 2.32 for function $F(ABCD)$:

F

CD \ AB	00	01	11	10
00	1	1	0	1
01	1	1	0	1
11	1	1	0	0
10	1	1	0	1

Fig. 2.32 Karnaugh map of F in Example 2.3.

According to its Karnaugh map, $F = \overline{A} + \overline{B}\,\overline{C} + \overline{B}\,\overline{D} = \overline{A} + \overline{B}(\overline{C} + \overline{D})$. Note that F, as a result of using a Karnaugh map, is expressed in product and sum terms of $\overline{A}$, $\overline{B}$, $\overline{C}$, and $\overline{D}$. The pull-up network can be readily created. However, we take a different route here to demonstrate the use of a Karnaugh map in the construction of CMOS logic circuits.

It can be readily seen that each square in the Karnaugh map that contains a 1 (1-square) indicates an input combination that turns on the pull-up network. Similarly, the squares that contain 0's (0-squares) indicate the input combinations that turn on the pull-down network. We replace the 0-squares and 1-squares in Fig. 2.32 with *On-*

squares and *Off*-squares, respectively, to create the Karnaugh map in Fig. 2.33 for the pull-down network.

Pull-down AB

		00	01	11	10
	00	*Off*	*Off*	*On*	*Off*
CD	01	*Off*	*Off*	*On*	*Off*
	11	*Off*	*Off*	*On*	*On*
	10	*Off*	*Off*	*On*	*Off*

Fig. 2.33 Karnaugh map for the pull-down network in Example 2.3.

The pull-down switch network for F is found by grouping the *On*-squares into prime implicants using the regular Karnaugh map operation:

$$\textbf{\textit{Pull-down}}: AB + ACD = A(B + CD). \tag{2.9}$$

Notice that in (2.9), variable A, which is common in both prime implicants, was extracted out as a common term to minimize the number of transistors in the pull-down network. The pull-down network of function F and its complementary pull-up network are shown in Fig. 2.35. The reader should study the relationship between the pull-down and pull-up networks to fully understand the application of Fig. 2.27.

Alternatively, we can create the following Karnaugh map for the pull-up network by replacing the 1- and 0-squares in the Karnaugh map for F by *On*- and *Off*-squares, respectively:

Pull-up AB

		00	01	11	10
	00	*On*	*On*	*Off*	*On*
CD	01	*On*	*On*	*Off*	*On*
	11	*On*	*On*	*Off*	*Off*
	10	*On*	*On*	*Off*	*On*

Fig. 2.34 Karnaugh map for the pull-up network in Example 2.3.

We can identify the prime implicants in Fig. 2.34 for the pull-up network as

$$\textbf{\textit{Pull-up}}: \overline{A} + \overline{B}\,\overline{C} + \overline{B}\,\overline{D} = \overline{A} + \overline{B}(\overline{C} + \overline{D}) \tag{2.10}$$

which, indeed, matches with the pull-up network shown in Fig. 2.35.

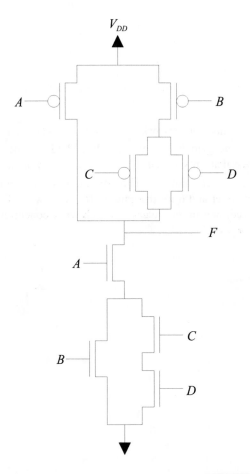

Fig. 2.35   Logic circuit for function $F = \overline{A(B+CD)}$.

The next example demonstrates the selection of an implementation from a number of functionally equivalent ones. The objective in this example is to minimize the number of transistors in the logic circuit.

Example 2.4

Design a CMOS complementary logic circuit for $F = A + BC$ and minimize the number of transistors used in the circuit.

The Karnaugh map of function F is shown in Fig. 2.36.

$$
\begin{array}{c}
\quad\quad\quad BC \\
F \\
\end{array}
$$

	BC 00	01	11	10
A 0	0	0	1	0
1	1	1	1	1

Fig. 2.36 Karnaugh map of F in Example 1.4.

The pull-down transistor network is found to be $\overline{A}(\overline{B}+\overline{C})$ by grouping the 0-squares. We follow the guideline of using only nMOS transistors in the pull-down network and assume that only signals A, B, and C are available. Inverters are then needed to produce $\overline{A}$, $\overline{B}$, and $\overline{C}$. This structure, which requires a total of 12 transistors (2 per inverter and 6 for the gate itself), is shown in Fig. 2.37. Two levels of delay are involved, one in the inverters which work concurrently, and one in the complementary gate itself.

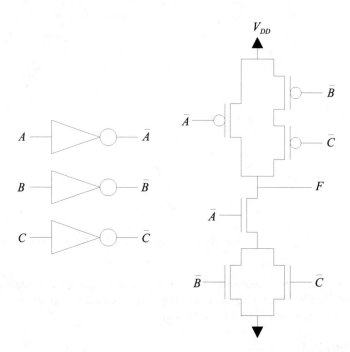

Fig. 2.37 Logic circuit designed for Example 2.4.

Alternatively, if we implement the logic function $Z = \overline{F} = \overline{A+BC}$, we will not need the inverters at the inputs. The pull-down network for Z is $A + BC$. We simply attach a single inverter at its output Z, and F is produced at the output of the inverter.

Fig. 2.38 shows this structure which requires only 8 transistors, a 33% saving from the circuit in Fig. 2.37. Note that both circuits have two levels of delay.

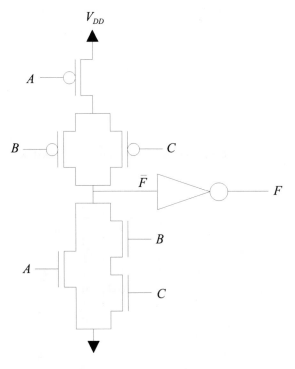

Fig. 2.38 Minimized logic circuit for Example 2.4.

2.6   Pass-Transistor/Transmission-Gate Logic

Complementary logic circuits are guaranteed to work. As long as we use only nMOS transistors in the pull-down network and pMOS transistors in the pull-up network, full logic swing (V_{DD} to V_{SS}) signals are produced at their outputs. However, if less than full logic swings can be accommodated in an application, or at least in a portion of it, pass-transistor logic may offer some saving in the number of transistors used in the circuit.

In a complementary logic circuit, the source of its output is always a constant logic signal: V_{DD} for the pull-up network and V_{SS} for the pull-down network. Pass-transistor logic also allows input signals and their complements to be passed on to the circuit output. This property in many cases can significantly simplify the circuit. We begin our discussion by contemplating the design of a 2-to-1 multiplexer (MUX), the symbol of which is shown in Fig. 2.39.

The operation of this 2-to-1 multiplexer is to selectively connect inputs i_0 or i_1 to output Z according to the logic signal on selection line S. Following the design method introduced in Section 2.5, we consider the multiplexer as a three-variable logic function $Z = f(S, i_0, i_1)$ and create the truth table shown in Fig. 2.40.

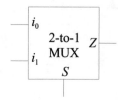

Fig. 2.39 Symbol of a 2-to-1 multiplexer (MUX).

S	i_0	i_1	Z
0	0	0	0
0	0	1	0
0	1	0	1
0	1	1	1
1	0	0	0
1	0	1	1
1	1	0	0
1	1	1	1

Fig. 2.40 Truth table of a 2-to-1 MUX.

This truth table in Fig. 2.40 is converted into the Karnaugh map in Fig. 2.41 to determine the pull-up and pull-down networks for Z:

Z	$i_0 i_1$			
	00	01	11	10
S 0	0	0	1	1
1	0	1	1	0

Fig. 2.41 Karnaugh map of a 2-to-1 MUX.

The Karnaugh map in Fig. 2.41 prescribes a logic function $Z = \overline{S}i_0 + Si_1$ for the multiplexer, of which the transistor schematic is shown in Fig. 2.42. This complementary logic circuit, which is our first design for the multiplexer, involves 14 transistors (2 per inverter and 8 for the complementary gate itself) and 2 levels of delay.

Alternatively, one can use the fact that $Z = i_0$ when $S = 0$, and $Z = i_1$ when $S = 1$ to rewrite the truth table of the 2-to-1 multiplexer as shown in Fig. 2.43. The reader should notice the similarity of the truth table in Fig. 2.43 to that of an inverter (Fig. 2.11). The only difference between them is that, for the multiplexer, in each row output Z is assigned an input variable (i_0 or i_1) instead of a constant (0 or 1) as in the case of an inverter. In a manner similar to the design of the inverter (Section 2.3), we create the pass-transistor logic circuit shown in Fig. 2.44 to selectively pass input signals to its output.

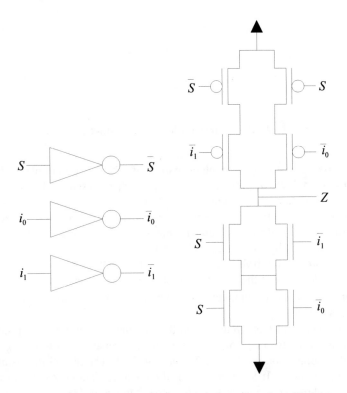

Fig. 2.42 Complementary logic circuit for a 2-to-1 multiplexer.

S	Z
0	i_0
1	i_1

Fig. 2.43 Modified truth table of a 2-to-1 multiplexer.

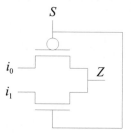

Fig. 2.44   Pass-transistor logic for a 2-to-1 multiplexer.

The pass-transistor logic implementation of the 2-to-1 multiplexer has only 2 transistors and 1 level of propagation delay, which represents a big saving over its counterpart complementary logic circuit (Fig. 2.42). In practice, many designers prefer to use only nMOS transistors in a pass-transistor logic circuit. The justification behind this design style is to avoid the big separation between nMOS and pMOS transistors required by the design rules (see Chapter 3). Another reason to prefer nMOS transistors over pMOS transistors is that nMOS transistors are inherently faster (see Chapter 4). If the pMOS transistor in Fig. 2.44 is replaced with an nMOS transistor, we would need an inverter to generate $\overline{S}$ for its control. This will bring the transistor count to 4.

The drawback of this circuit lies in the fact that output Z cannot produce a full logic swing (V_{DD} to V_{SS}) for certain input combinations. Recall the situations discussed in connection with Fig. 2.20 and Fig. 2.21. The nMOS transistor degrades a 1 passing through it. If $S = 1$ and $i_1 = 1$, the voltage at output Z cannot reach above ($V_{DD} - V_{tn}$) and becomes a weak 1. On the other hand, a pMOS transistor degrades a 0 passing through it, so output Z can only go down to $|V_{tp}|$ — a weak 0, if $S = 0$ and $i_0 = 0$.

In many applications, however, pass-transistor logic is accepted as a valid alternative to complementary logic. Special techniques have been developed to deal with its less than perfect logic swings and other limitations. For example, the weak signals can be restored by a following complementary logic stage. Fig. 2.45 shows an approach to restore the logic level at the output of a pass-transistor (B). The pMOS transistor is turned on whenever Z is 0, which reflects the fact that A is a 1; so B

should also be a 1. The pMOS transistor pulls B up to V_{DD} and thus restores the logic level.

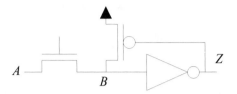

Fig. 2.45   Trickle transistor used to restore logic level.

Alternatively, the transistors in a pass-transistor logic circuit can be replaced with transmission-gates (t-gates). A t-gate is built by connecting a pMOS transistor and an nMOS transistor in parallel. The structure and the symbol of the t-gate are shown in Fig. 2.46.

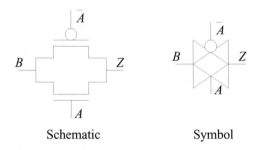

Schematic   Symbol

Fig. 2.46 Transmission-gate (t-gate) structure and its symbol.

A t-gate provides a non-clipping switch for both 0 and 1. Complementary control signals (A and $\overline{A}$) are applied to the pMOS and nMOS transistors in a t-gate so both transistors turn on and off simultaneously. A t-gate does not cause a voltage drop to either the 1 or the 0 passing through it. The nMOS transistor, when turned on, provides the path for the 0. The conducting pMOS transistor passes the 1. The nMOS transistor provides a signal path between B and Z when $B = 0$. The pMOS transistor, on the other hand, provides the signal path between B and Z when $B = 1$.

Fig. 2.47 shows the result of converting the 2-to-1 multiplexer of Fig. 2.44 into a t-gate logic. An additional inverter (not shown) is needed to generate control signal $\overline{S}$. The transistor count of this circuit is six (inverter included), which is three times that of the pass-transistor multiplexer.

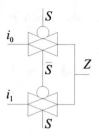

Fig. 2.47 Transmission-gate logic for a 2-to-1 multiplexer.

The 2-to-1 multiplexer (MUX) can be used as a building block to create multiplexers with more selectable inputs. Fig. 2.48 and Fig. 2.49 show the construction of a 4-to-1 multiplexer and an 8-to-1 multiplexer, respectively. The application of the VLSI hierarchical design methodology, which emphasizes the partitioning of a large system into manageable building blocks, is apparent in these two examples.

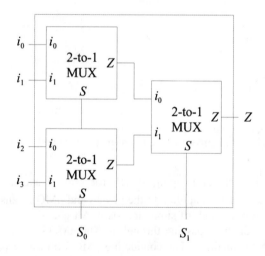

Fig. 2.48 4-to-1 multiplexer.

Multiplexers can be used as configurable logic components to implement desired logic functions. For instance, the 8-to-1 multiplexer shown in Fig. 2.49 provides a configurable structure for the implementation of any three-variable logic functions.

Fig. 2.50 demonstrates the use of an 8-to-1 multiplexer to implement the logic function $F = A + BC$ given in Example 2.4. Variables A, B, and C are applied to the selection inputs S_2, S_1, and S_0, respectively. The truth table is implemented by hardwiring each of the multiplexer inputs $i_0 - i_7$ to either 1 or 0, according to function F. This implementation is straightforward. No optimization procedure is involved.

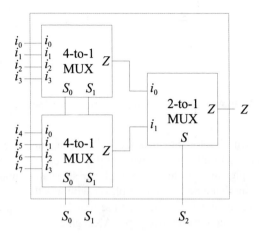

Fig. 2.49 8-to-1 multiplexer.

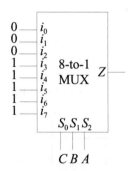

Fig. 2.50 Multiplexer hard-wired to implement $F = A + BC$.

Multiplexer-based logic has been extensively used to build the configurable logic blocks of field programmable gate arrays (FPGAs). Fig. 2.51 shows a register which stores the outputs of a desired logic function. The outputs of this register are applied to the selectable inputs of a multiplexer. The three input variables of the logic function are applied to the multiplexer selection lines $(S_2 S_1 S_0)$ to select one of the input lines $(i_7 - i_0)$ as the output Z.

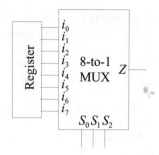

Fig. 2.51 A multiplexer-based configurable logic block.

A multiplexer with n selection lines ($S_0 - S_{n-1}$) can implement a general logic function of n variables. In addition, if we apply some of the n variables and their complements of a logic function to the multiplexer selectable input lines, the logic function can also be implemented by a multiplexer with less than n selection lines.

The following example demonstrates the use of a 2-to-1 multiplexer, which has a single selection line, to implement a two-input AND gate.

Example 2.5

Use a 2-to-1 multiplexer to implement a two-input AND gate ($F = AB$), the logic symbol of which is shown in Fig. 2.52.

Fig. 2.52 Logic symbol of a two-input AND gate.

A	B	F
0	0	0
0	1	0
1	0	0
1	1	1

Fig. 2.53 Truth table of $F = AB$ in Example 1.5.

Since a 2-to-1 multiplexer has only one selection line S, we arbitrarily choose to assign input A to S. According to the truth table of F in Fig. 2.53, if $A = 0$, F equals 0 regardless of B's value. On the other hand, when $A = 1$, F is an identical function of B (i.e., $F = B$). The logic function $F = AB$ can then be rewritten as $F = \overline{A}0 + AB$ and implemented with a 2-to-1 multiplexer as shown in Fig. 2.54.

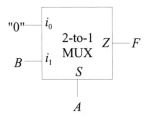

Fig. 2.54 2-to-1 multiplexer implementing a two-input AND gate.

Example 2.6

Use a 4-to-1 multiplexer to implement $F = \overline{A}(B + \overline{C}) + A\overline{B}C$.

F is a logic function of 3 variables but a 4-to-1 multiplexer has only 2 selection lines. We first assign inputs A and B to the selection lines S_1 and S_0, respectively. The relationship between F and the third input C in the truth table (Fig. 2.55) is then analyzed.

A	B	C	F
0	0	0	1
0	0	1	0
0	1	0	1
0	1	1	1
1	0	0	0
1	0	1	1
1	1	0	0
1	1	1	0

Fig. 2.55 Truth table of $F = \overline{A}(B + \overline{C}) + A\overline{B}C$.

It is readily seen from the first two row of the above truth table that when $AB =$ 00, the value of F depends on C (i.e., $F = \overline{C}$). The dependency of F on C for the other combinations of AB can be found in a similar manner. The truth table in Fig. 2.56 summarizes these relations.

A	B	F
0	0	$\overline{C}$
0	1	1
1	0	C
1	1	0

Fig. 2.56 Modified truth table of $F = \overline{A}(B + \overline{C}) + A\overline{B}C$ in Example 1.6.

Fig. 2.57 shows the use of a 4-to-1 multiplexer to implement this function.

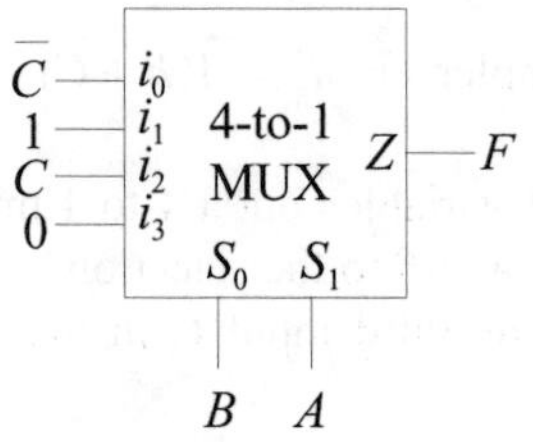

Fig. 2.57 $F = \overline{A}(B + \overline{C}) + A\overline{B}C$ implemented with a 4-to-1 multiplexer.

Tri-state logic gates can be formed by attaching a t-gate to the output of a regular logic gate. Fig. 2.58 shows the structure of a tri-state inverter. When the t-gate is turned on, the circuit functions as an inverter. If the t-gate is turned off, Z is at a high impedance state. Other tri-state logic gates can be built in a similar manner.

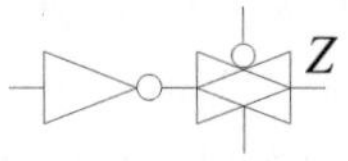

Fig. 2.58 Tri-state inverter.

2.7 Summary

Among different CMOS logic structures, the complementary logic circuit is most reliable. Both nMOS and pMOS transistors can be modeled as switches that are turned on by digital signals 1 and 0, respectively. The simple switch model does not have the capability to do delay and power estimation, but is adequate for the design and analysis of CMOS logic circuits.

Karnaugh maps provide a manual tool for the design of small complementary logic circuits. The pull-up network of a logic circuit, which connects the output to V_{DD} (1) when turned on, has a structure determined by the prime implicants found by grouping the "1-squares" in the Karnaugh map. On the other hand, the pull-down network, which connects the output to V_{SS} (0) when turned on, can be found by identifying the prime implicants of the "0-squares" in the Karnaugh map. The complementary relationship between the pull-up and pull-down networks allows one network to be found by complementing the structure of the other.

Three tangible criteria are commonly used to evaluate and select an implementation from functionally identical but structurally different circuits: area, delay, and power dissipation. The exact area of a circuit is only available after the circuit is laid out. The number of transistors in a circuit is thus commonly used to estimate its area.

The number of logic stages indicates the levels of delay; so it can be used as an estimate of circuit speed. Because the complementary logic circuit isolates the power source from ground when the circuit is in a steady state, its static power dissipation can normally be ignored.

The switch networks in a complementary logic circuit selectively connect its output to constant logic signals (1 in the pull-up network and 0 in the pull-down network). A pass-transistor network removes this restriction and allows the output to be connected to 0, 1, an input, or an input's complement. In many cases, a logic function can be implemented with fewer transistors by a pass-transistor network than a complementary one. Instead of pass-transistors, transmission gates may be used to remedy the weak signal drawback of a pass-transistor circuit.

2.8   To Probe Further

MOSFET and CMOS electronics:

- S. Sedra and K. C. Smith, *Microelectronic Circuits, 4th Edition*, Oxford University Press, 1998.

- K. Lee, M. Shur, T. A. Fjeldly, and T. Ytterdal, *Semiconductor Device Modeling for VLSI*, Prentice-Hall, 1993.

SPICE:

- G. W. Roberts and A. S. Sedra, *SPICE*, Oxford University Press, 1997.

Digital design:

- J. F. Wakerly, *Digital Design Principles and Practices*, 2nd Edition., Prentice-Hall, 1994.

- V. P. Nelson, H. T. Nagle, B. D. Carroll, and J. D. Irwin, *Digital Logic Circuit Analysis & Design*, Prentice-Hall, 1995.

Karnaugh map:

- M. Karnaugh, "The map method of combinational logic circuits," *AIEE Comm. Electronics*, November 1953, pp. 593-599.

Quine-McCluskey logic minimization:

- E. J. McCluskey, Jr., *Introduction to the Theory of Switching Circuits*, McGraw-Hill, 1965.

- E. J. McCluskey, Jr., "Minimization of Boolean functions," *Bell System Tech. J.*, November 1956, pp. 1417-1444.

- W. V. Quine, "The problem of simplifying truth functions," *Amer. Math. Monthly*, vol. 59, 1952, pp. 521-531.

2.9   Problems

The following logic functions are to be used in Problems 2.1 to 2.7:

(i) $F = A \oplus B$

(ii) $G = \overline{A \oplus B}$

(iii) $H = A\overline{B}\ \overline{C}D + \overline{A}(BC\overline{D} + \overline{B}\ \overline{C})$

(iv) $K = (\overline{A} + B)(C + D + E)(\overline{B} + D)$

(v) $L = AB + CD$

(vi) $M = \overline{AB + CD}$

(vii) $N = \overline{AB} + C\overline{D}$

(viii) $P = \overline{\overline{AB} + C\overline{D}}$

2.1   Assume that only signals A, B, C, and D are available. Implement each of the given logic functions as complementary logic circuits. Minimize the number of transistors in each circuit.

2.2   Assume that only signals A, B, C, and D are available. Implement each of the given logic functions as complementary logic circuits. Minimize the number of gate levels in each circuit.

2.3   Given a logic function Z, there are two ways to implement it as a complementary logic circuit. A complementary logic gate can be developed for Z. Alternatively, a complementary logic gate can be developed for $\bar{Z}$ with an inverter attached to its output to produce Z. In either case, inverters may be needed for the gate inputs to generate required input signals.

Design a method that inspects the Boolean expression of a logic function and determines, for the objective of minimizing the number of transistors, the appropriate way to implement the logic function. Test the validity of your method with the given logic functions.

2.4   Implement the given logic functions as pass-transistor logic circuits. Compare your designs (number of transistors, delay levels, design complexities, etc.) with those developed in Problems 2.1 and 2.2.

2.5   Implement the given logic functions with an 8-to-1 multiplexer and inverters. Compare your designs (number of transistors, delay levels, design complexities, etc.) with those developed in Problems 2.1 and 2.2.

2.6   What is the output voltage (v_o) in terms of V_{DD} and V_{tn} of the circuit in Fig. 2.59?

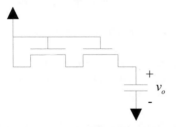

Fig. 2.59 Circuit for Problem 2.6.

2.7   What is the output voltage (v_o) in terms of V_{DD} and V_{tn} of the circuit in Fig. 2.60?

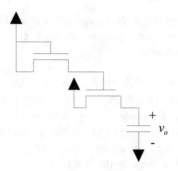

Fig. 2.60 Circuit for Problem 2.7.

2.8   Consider the logic diagram given in Fig. 2.61.

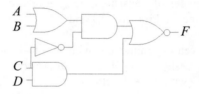

Fig. 2.61 Circuit for Problem 2.8.

(a) Implement F as a complementary logic function. Minimize the number of transistors.
(b) Implement F as a complementary logic function. Minimize the number of delay levels.
(c) Use a 16-to-1 multiplexer to implement F.
(d) Use only standard-cells (inverters, two-input NAND gates, and two-input NOR gates) to implement F.

2.9   Assume MOSFET transistors are perfect switches and identify the logic function for the circuit in Fig. 2.62.

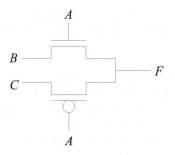

Fig. 2.62 Circuit for Problem 2.9.

2.10   Explain why NAND and NOR gates are preferred to AND and OR gates in CMOS technology.

2.11   Is the circuit shown in Fig. 2.63 a valid logic function? If yes, what is the logic function F? Note that the structures of its pull-up and pull-down networks do not follow the rules given in Fig. 2.27.

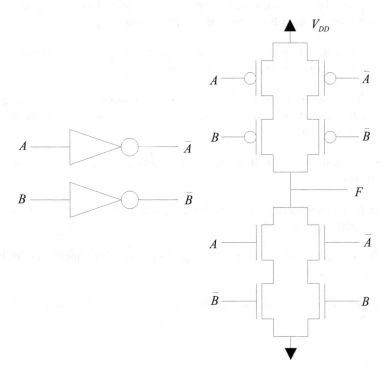

Fig. 2.63 Circuit for Problem 2.11.

2.12   Design a minimized complementary logic circuit to implement a majority voter. This voter should have three inputs and produce a 1 when two or more of its inputs are 1's.

2.13   Design a minimized complementary logic circuit for a two-bit binary adder/subtractor. The circuit accepts two binary numbers (a_1a_0 and b_1b_0) and one control signal (s). The output of this circuit has three bits ($z_2z_1z_0$). When s = 1, $z_2z_1z_0 = a_1a_0 - b_1b_0$, otherwise $z_2z_1z_0 = a_1a_0 + b_1b_0$. Note that z_2 indicates a "carry" in addition and a "borrow" in subtraction.

2.14   Use three 16-to-1 multiplexers as the major building blocks to implement the adder/subtractor described in Problem 2.13. Minimize your design.

2.15   Use three 8-to-1 multiplexers as the major building blocks to implement the adder/subtractor described in Problem 2.13. Minimize your design.

2.16   Design a minimized complementary logic circuit that accepts a 4-bit unsigned binary number and outputs a 1 if and only if the number is $(7)_{10}$ or less.

2.17   Use one 16-to-1 multiplexer as the major building block to implement the circuit described in Problem 2.16. Minimize your design.

2.18   Use one 8-to-1 multiplexer as the major building block to implement the circuit described in Problem 2.16. Minimize your design.

2.19   Design a minimized CMOS comparator that compares two 2-bit numbers A and B. The comparator has a 2-bit output, which is 00 if $A = B$, 01 if $A > B$, and 10 if $A < B$.

2.20   Design a minimized CMOS circuit that produces the 2's complement of a 4-bit number.

2.21   Design a minimized CMOS combinational multiplier that multiplies two 2-bit numbers (a_1a_0, and b_1b_0).

2.22   Design a non-clipping 2-input AND gate with only five transistors (pMOS and nMOS).

2.23   Design a non-clipping 2-input OR gate with only five transistors (pMOS and nMOS).

2.24 Identify the function of the circuit shown in Fig. 2.64.

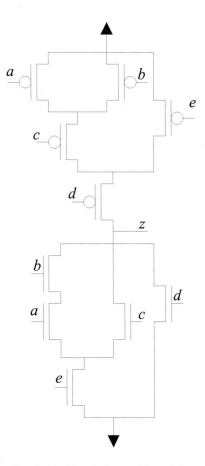

Fig. 2.64 Circuit for Problem 2.24.

2.25 Design a CMOS circuit to implement the truth table shown in Fig. 2.65. This circuit has 4 data inputs (a_0-a_3), 4 control inputs (c_0-c_3), and 4 data outputs (z_0-z_3).

c_3	c_2	c_1	c_0	z_3	z_2	z_1	z_0
0	1	1	1	a_3	a_2	a_1	a_0
1	0	1	1	a_2	a_1	a_0	a_3
1	1	0	1	a_1	a_0	a_3	a_2
1	1	1	0	a_0	a_3	a_2	a_1

Fig. 2.65 Truth table for Problem 2.25.

Chapter 3 IC Layout and Fabrication

Sand turned into ICs ...

As we have explained in Chapter 1, a significant amount of effort has been applied to decouple the design of an IC from its fabrication. For example, one significant milestone in the development of VLSI technology during the 70's was the creation of design rules to specify fabrication limitations as geometric constraints on the layout. The concept of IC design rules allows design activities to be carried out separately from the fabrication process. Furthermore, computer-aided design (CAD) tools have been created to synthesize application-specific ICs (ASICs) from their behavioral representations. Despite these and other efforts, an understanding of the silicon IC fabrication technology is invaluable to the appreciation of many VLSI research and design issues.

This chapter first introduces the necessary background of silicon IC fabrication technology. Readers who desire to learn more about the subject can consult the references listed at the end of this chapter. We then proceed to discuss the creation of an IC physical layout, which is a geometric representation of the transistors in a circuit and their interconnections. The relationship between a layout and its corresponding IC physical structure is explained.

3.1 CMOS IC Fabrication

Despite that germanium (Ge) and gallium arsenic (GaAs) are also used to fabricate integrated circuits, silicon (Si) is currently the most popular IC material. Many researchers in the IC industry predict that the popularity of silicon will sustain into the foreseeable future.

Silicon, which occurs naturally in the form of sand, makes up 28% of the Earth's crust. In addition to being an abundant material, silicon has a number of advantages over other materials for IC fabrication. First, silicon has the physical properties needed for creating transistors with good characteristics. Silicon wafers, on which CMOS integrated circuits are built, can be produced using relatively simple and inexpensive techniques. Silicon dioxide (SiO_2) can be easily formed on the surface of a silicon wafer as a diffusion barrier to selectively alter the electrical properties in the wafer.

Semiconductor manufacturers usually purchase ready-to-use silicon wafers from wafer foundries. Fig. 3.1 illustrates a typical silicon wafer production process. The first step is to grow a single-crystal silicon ingot. Highly purified polycrystalline silicon is molten. A carefully controlled amount of impurities is then added to produce required electrical properties. A seed crystal is dipped into the molten silicon to initiate the growth of a single crystal. The seed is slowly rotated and pulled vertically out of the melt. The molten silicon attaches to the seed and forms an ingot that assumes the crystal orientation of the seed. The diameter of the ingot is determined by the rate of pulling and rotating the seed. The silicon ingot is then sawed into wafers 10 to 30 cm in diameter and 400 to 600 μm thick. At least one of the surfaces of the wafer is polished to a mirror finish free of scratches. The electrical properties of the wafer are determined by the crystal orientation, and the concentration and type of impurities added to the molten silicon.

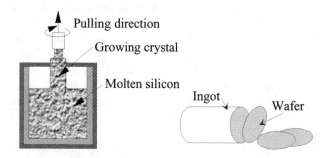

Fig. 3.1 Silicon wafer manufacturing process.

The creation of a circuit on a silicon wafer involves two major types of operations: doping impurities into selected wafer regions to change electrical properties and depositing patterned materials on the wafer surface.

Fig. 3.2 shows the steps of a selective doping. A layer of silicon dioxide (oxide layer) is grown by oxidizing ($SiO_2 \leftarrow Si + O_2$) the surface of the wafer (Fig. 3.2a). The oxide layer is then covered with a layer of photosensitive material called photoresist (Fig. 3.2b). The photoresist-coated wafer is exposed to UV light or electron beam through a mask that defines a pattern (Fig. 3.2c). The photoresist residing in exposed regions softens. The softened photoresist is removed using a chemical developer. This type of photoresist, which protects wafer regions corresponding to the opaque regions on a mask, is called a positive photoresist. In contrast, a negative photoresist allows the unexposed regions to be dissolved.

The wafer is treated in a chemical that removes unprotected silicon dioxide. The photoresist-covered regions are inactive to the chemical. Windows, defined by the pattern on the mask, are thus opened in the oxide layer (Fig. 3.2d). Impurity atoms, or dopants, are then introduced to the wafer surface but only accepted by the patterned region with no oxide on top (Fig. 3.2e). A diffusion area is thus created by the dopants moved through the crystal lattice. The depth of the diffusion area is determined by the temperature and time of doping. The most common p-type dopant is boron (B). Phosphorus (P) and arsenic (As) are commonly used as n-type dopants.

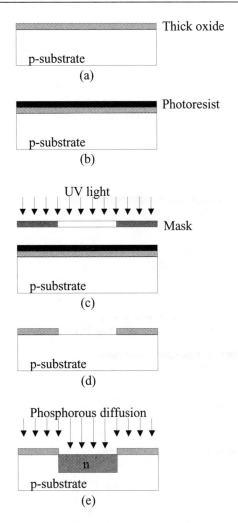

Fig. 3.2 Doping to a selected wafer region.

Another important process is to build a layer of patterned material such as aluminum or polysilicon on top of the wafer. Polysilicon are used to form transistor gates and short distance interconnections. Metal (aluminum) is used for the distribution of power and signals.[1] Fig. 3.3 illustrates this process. Multiple layers of material, insulated from each other by silicon dioxide, can be formed on a wafer surface.

---

[1] In 1997, IBM scientists developed a process to use copper instead of aluminum as conducting material. Copper has better conductivity than aluminum.

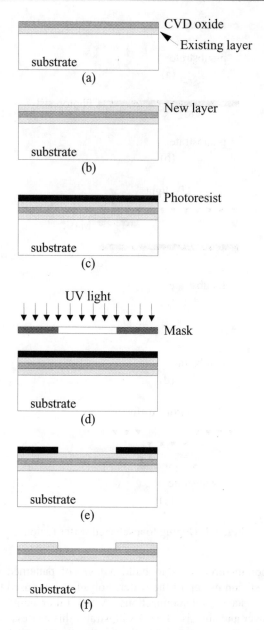

Fig. 3.3 Creating layers on top of the wafer.

Unlike on the bare surface of a wafer, silicon dioxide cannot be formed on top of a previously added layer by oxidizing the wafer. A process called chemical vapor deposition (CVD) is used instead to form a solid layer on the wafer top by chemically reacting gases or vapors. For example, the insulating oxide layer is formed by depositing silane gas and oxygen (Fig. 3.3a).

Alternatively, if CVD is performed with silane gas alone at less than $1000°C$, a layer of polycrystalline silicon (polysilicon) is formed. Polysilicon itself is not a good conductor so it is heavily doped to provide an adequate conductivity. Alternatively, a metal layer can be formed by heating aluminum in vacuum until it vapors and deposits in the silicon surface (Fig. 3.3b). A photoresist layer exposed through a mask is developed into a protective coating of the material deposited on the wafer so that unwanted regions are etched away by chemicals (Fig. 3.3c - f).

These two operations are carried out multiple times to process a wafer into an array of chips, which are then sawed out into chips and packaged. A simplified CMOS process is explained with an example of creating an inverter. CMOS circuits require both pMOS and nMOS transistors to be built on the same substrate. Depending on the type of wafer, CMOS circuits can be built with n-well, p-well, or twin-well processes. Alternatively, in the silicon on insulator (SOI) technology, nMOS and pMOS transistors are built on a neutral substrate (i.e., insulator).

The foundation of an n-well process is a p-type wafer. A region on the p-type wafer has to be doped and converted into n-type for the pMOS transistors. Alternatively, an n-type wafer with p-type areas created for the nMOS transistors is used in a p-well process. In a twin-well process, both well types are formed in a relatively neutral wafer so it is most flexible but more complicated. Currently the n-well process is more popular because it is compatible with the BiCMOS process that incorporates BJTs in a MOSFET circuit to enhance its speed and current capability.

A typical CMOS process requires the use of ten or more masks, depending on the number of polysilicon and metal layers needed. In the following discussion, we present the fabrication of an inverter by showing the results of major processing steps.

Fig. 3.4 provides four different views of an inverter. Fig. 3.4a shows the transistor schematic that we have developed in Chapter 2. Fig. 3.4b shows a layout, which is a composite view of the masks stacked together. Fig. 3.4c provides the top view[2] of the inverter manufactured according to the layout. Fig. 3.4d shows the cross sectional view of the same inverter. We now proceed to look at the fabrication of such an inverter.

---

[2] This top view is only illustrative since we pretend that we can see through the polysilicon and metal layers.

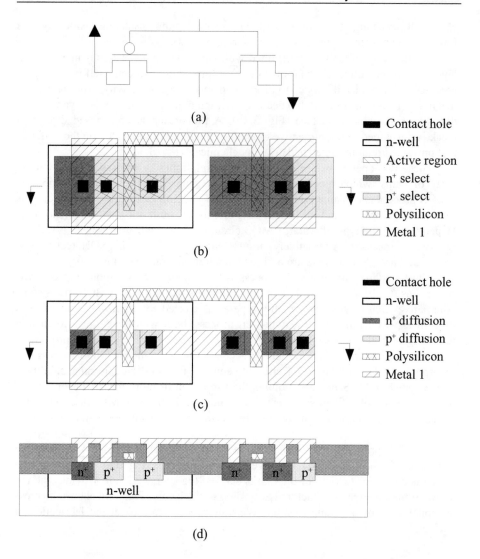

Fig. 3.4 CMOS inverter: (a) transistor schematic; (b) composite layout;
(c) top view; (d) cross sectional view.

The process of creating an n-well inverter (or in general, any n-well CMOS circuits) can be divided into three stages. In the first stage, n-wells are formed and active regions are defined for the transistors (Fig. 3.5). The transistors are created in the second stage (Fig. 3.6). In the third stage, the metal interconnections are formed (Fig. 3.7).

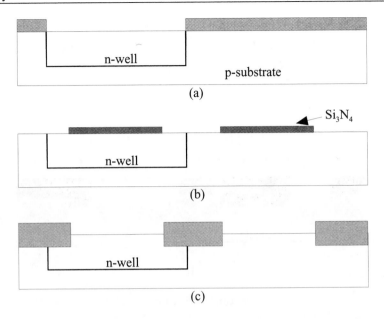

Fig. 3.5 Creating an n-well for CMOS circuits.

Fig. 3.5 illustrates that n-wells, one for each group of p-MOS transistors, are created with doping (Fig. 3.5a). Silicon nitride (Si_3N_4) is deposited on the substrate and patterned with an active region mask to cover active areas defined for transistors (Fig. 3.5b). A thick field oxide is grown everywhere on top of the substrate except those areas covered by nitride (Fig. 3.5c). Fig. 3.5c shows the thick field oxide created to isolate the pMOS and nMOS transistors to prevent a problem called latch-up (to be described in Chapter 4) from occurring.

A misalignment between the transistor gate and channel will cause the transistor to fail. A solution to this problem is the use of a technique called the self-aligned process to create the transistors. In this technique, the polysilicon gate is created before the diffusion of the drain and source. The polysilicon gate is then used as a barrier when the doping is performed on the entire active region so that a channel region is defined with the gate itself. This guarantees the alignment between the gate and the channel.

The self-aligned process of creating transistors is illustrated in Fig. 3.6. A layer of gate oxide is grown on the wafer. A layer of heavily doped polysilicon is deposited and patterned with a mask to form the gates and short interconnections (Fig. 3.6a). A layer of photoresist is patterned with an n^+-select mask to allow the sources and drains of nMOS transistors to be doped with phosphorous (Fig. 3.6b). Notice the n^+ well contact of the n-well prepared for pMOS transistors is also created in the same step. Similarly, the pMOS transistors and the substrate contact of the wafer are created by doping their drains and sources with boron (Fig. 3.6c).

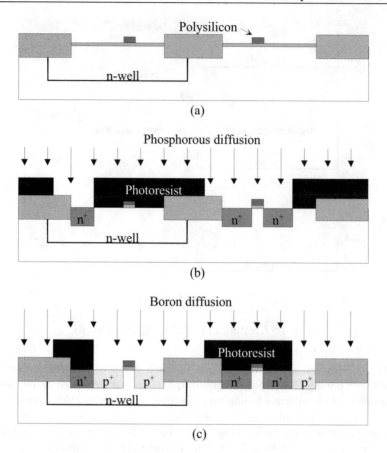

Fig. 3.6 Self-aligned process of creating transistors for the inverter.

   The metallization process is shown in Fig. 3.7. A thick layer of oxide is
deposited over the entire wafer. Contact cuts, which provide the paths for
interconnecting different layers, are defined with a mask (Fig. 3.7a). These contact
cuts are often filled with tungsten to ensure connectivity. A thin layer of aluminum is
evaporated onto the substrate, patterned, and etched to form the interconnections
(Fig. 3.7b). A layer of thick CVD oxide (glass) is usually deposited on the finished
wafer as a protective layer (not shown).

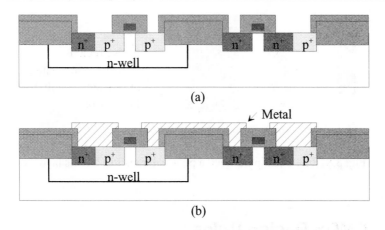

(a)

(b)

Fig. 3.7 Metallization process of a CMOS inverter.

Refer back to Fig. 3.4b, which shows two masks are required to form the diffusion regions (i.e., source and drain) of a transistor. An active region mask is first used to define the dimensions of the transistors. The doping is then actually done with a select mask with openings that extend beyond the active regions. This way we can ensure that the transistors are formed correctly. This can be seen by observing the difference between the composite layout diagram (Fig. 3.4b) and the top view (Fig. 3.4c). The active region masks define where the diffusion actually takes place although the n^+-select and p^+-select masks cover areas extended beyond the transistors. Also, there is no diffusion underneath the polysilicon gates (i.e., the channels) although both the active and select masks include the channels to conform to the self-aligned process.

It is often more convenient for a layout designer to think and work directly with the transistor source/drain dimensions which are defined by the active region mask. The symbolic layout style pioneered by the layout tool *magic* developed at University of California, Berkeley allows the designer to create a transistor by specifying only its active regions (n^+ and p^+ diffusions) and its gate.[3] The select mask and well mask for the transistors can be automatically generated according to the design rules.

Fig. 3.8 demonstrates the relationship between a physical nMOS transistor and its symbolic and composite layouts. Fig. 3.8a shows the top view of this nMOS transistor. As noted, physically there is no diffusion in the channel underneath the polysilicon gate. Fig. 3.8b is a symbolic transistor layout that specifies the n^+ diffusion (active) areas and the gate. This symbolic layout can be converted into the composite mask layout shown in Fig. 3.8c, which includes the active region, n^+-select, and polysilicon masks.

---

[3] *Magic* is a full-custom symbolic layout tool with interactive design rule checking and can be downloaded from http://www.research.digital.com/wrl/projects/magic.

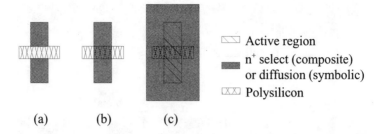

Fig. 3.8 Relationship between a physical transistor and its layouts.

3.2 CMOS Design Rules

We have shown in Fig. 3.4 that a layout resembles the top view of the IC. The layout design is thus 2-dimensional from the viewpoint of an IC designer. In fact, designers normally do not control the depth dimension of an IC. The depth of a transistor source or the thickness of a metal wire is determined by the fabrication process. The layout designer only decides the dimensions and locations of the transistors and interconnects them into the target circuit. The minimum feature size of a technology is typically denoted by the narrowest width of a polysilicon wire that it can produce. For instance, if the narrowest polysilicon wire in a technology is 1 μm wide, it is called a 1 μm technology.

Design rules include width rules and spacing rules. They capture the physical limitations of a fabrication process to ensure that a design that conforms to the design rules can be successfully fabricated. Therefore, design rules release IC designers from the details of fabrication so they can concentrate on the design instead. Note that the successful fabrication of an IC does not automatically imply that it meets the design criteria. It simply means that the transistors and their interconnections specified in the layout are operational.

Mead and Conway developed a set of simplified design rules now known as the scalable λ-based design rules, which are valid for a range of fabrication technologies.[4] In these rules, the minimum feature size of a technology is characterized as 2 λ. All width and spacing rules are specified in terms of the parameter λ. Suppose we have design rules that call for a minimum width of 2 λ and a minimum spacing of 3 λ. If we select a 2 μm technology (i.e., λ = 1 μm), the above rules are translated to a minimum width of 2 μm and a minimum spacing of 3 μm. On the other hand, if a 1 μm technology (i.e., λ = 0.5 μm) is selected, then the same width and spacing rules are now specified as 1 μm and 1.5 μm, respectively.

---

[4] C. A., Mead and L. A. Conway, *Introduction to VLSI Systems*. Addison-Wesley, 1980.

Different fabrication technologies apparently require different design rules. The scalable λ-based design rules are possible because they are created to be sufficiently conservative for a range of fabrication technologies.

The readers should be aware of the drawbacks of scalable design rules: the layouts produced according to scalable design rules are often larger than necessary. In order to accommodate a range of fabrication processes, these rules are naturally conservative. For example, a minimum spacing of 4 μm may be ordered although in reality 3.2 μm is sufficient. This limitation is especially critical in a full-custom design in which an optimized layout area is desired.

The usefulness of scalable λ-based design rules is limited for sub-micron technologies. We demonstrate the use of scalable λ-based design rules to create layouts since these rules are easier to learn and they are popular in MOSIS supported projects.[5] Technology specific design rules should be used in performance critical designs.

MOSIS Design Rules

MOSIS publishes CMOS design rules for processes with a range of feature sizes. Currently, MOSIS design rules can be found in their web page: http://www.mosis.org. The reader should download the design rules for a target process as a reference. We use a subset of MOSIS' scalable CMOS (SCMOS) design rules to provide an overview of their use. This subset covers the rules for one polysilicon layer, two metal layers, and enhancement-mode nMOS and pMOS devices. SCMOS compatible fabrication processes with additional polysilicon and metal layers are available through MOSIS. Design rules revised for better fit to sub-micron (feature size < 1 μm) and deep sub-micron (feature size < 0.25 μm) processes are also available.

We concentrate on a subset of the SCMOS rules used in a symbolic layout process that was popularized by the computer-aided layout tool *magic*. In order to maintain the tool independence of the following description, the discrepancy between the symbolic and physical layouts will be pointed out when the situation warrants it. The legends used in the illustrations of design rules are given in Fig. 3.9.

Polysilicon is used both as transistor gates (gate poly) and as short distance interconnections (field poly). Fig. 3.10 shows that the minimum width of a polysilicon wire is 2 λ (the minimum feature size) and two unrelated polysilicon wires must be separated by at least 2 λ.

---

[5] MOSIS provides VLSI circuit prototyping services for academic, research, and commercial institutes. See http://www.mosis.org.

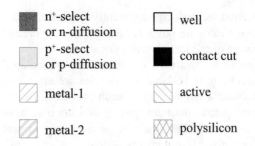

Fig. 3.9   Legends for different IC mask layers.

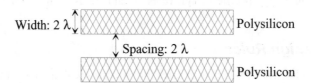

Fig. 3.10   Polysilicon design rules.

The following rules for diffusion regions are valid for both n-type and p-type diffusion regions. Fig. 3.11 shows that the minimum width of a diffusion region is 3 λ. Two unrelated diffusion regions of the same type have to be separated by at least 3 λ.

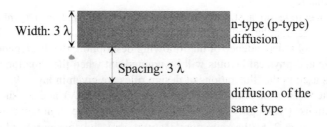

Fig. 3.11   Diffusion (n-type or p-type) design rules.

The largest spacing rule in the scalable λ-based design rules is the minimum distance between diffusion regions of different types. They have to be separated by at least 10 λ. The reason behind this large spacing is to avoid the latch-up problem, which can cause a CMOS circuit to malfunction. Latch-up will be discussed in Chapter 4.

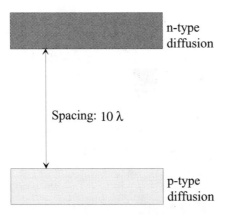

Fig. 3.12 Spacing between different types of diffusion.

In a symbolic layout, transistors are formed by crossing polysilicon wires and diffusion regions. The action of running a polysilicon wire across a p-type diffusion indicates a pMOS transistor. An nMOS transistor is specified by drawing a polysilicon wire across an n-type diffusion. Only the active diffusion regions and their types have to be specified in a symbolic layout. Both the select mask and well mask can be generated, manually or automatically, at a later stage.

Fig. 3.13 illustrates the symbolic transistor design rules. The layout of two serially connected pMOS transistors is demonstrated.[6] The formation of a connection between the first metal (metal-1) layer and the diffusion layer, called a diffusion-contact, is also shown. The channel area is defined by the overlapping area of the polysilicon wire and the diffusion. The polysilicon wire must extend beyond the diffusion that it is crossing for at least 2 λ. A field poly wire must be separated from a diffusion region for at least 1 λ. The diffusion region of a transistor must be extended from its gate for at least 3 λ. The minimum size of a diffusion-contact is 4 λ × 4 λ. Its details will be further discussed in a moment. A channel must be separated from a diffusion-contact by at least 1 λ.

Fig. 3.13 also illustrates the application of the polysilicon wire separation rule (2 λ) in forming the two serially connected transistors. The two transistor gates are separated by 2 λ. The diffusion area between these transistor gates serves as the drain of the transistor at the top. At the same time, it is also the source of the lower transistor. In other words, the diffusion region is used in this layout example as a local interconnect between transistors.

---

[6] Identical design rules apply to the design of nMOS transistors.

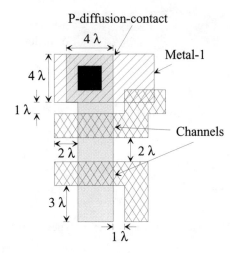

Fig. 3.13 Transistor (symbolic) design rules.

A composite layout generated from the symbolic layout in Fig. 3.13 is shown in Fig. 3.14. The select area must be extended a minimum of 2 λ beyond the diffusion (active) area it covers. Note that a p-select is used to create the p-diffusion areas for a pMOS transistor and an n-select is needed to create the n-diffusion areas for an nMOS transistor. In addition, a p-diffusion area must be built in an n-well that must be extended at least 5 λ beyond the edges of its enclosed diffusion area.

Fig. 3.15 shows that metal-1 (the first metal layer) wires are required to have a minimum width of 3 λ. Unrelated metal-1 wires are separated by at least 3 λ.

Fig. 3.16 shows that the minimum width of a metal-2 (the second metal layer) wire is the same as that of a metal-1 wire (3 λ), but the spacing between unrelated metal-2 wires is at least 4 λ.

Fig. 3.17 illustrates the design rules for interlayer contacts. An interlayer contact can be made from a layer to the layer above or below it. Contacts are allowed between substrate and metal-1, well and metal-1, n-diffusion and metal-1, p-diffusion and metal-1, polysilicon and metal-1, metal-1 and metal-2. All contacts are made by overlapping the two layers to be connected. A contact-cut, which is an opening in the silicon dioxide between the two layers to be connected, is provided at the center of the overlapped area. Metal such as tungsten is deposited in the contact-cut to form a contact between the top and bottom layers.

The two contacting layers must overlap for a minimum area of 4 λ × 4 λ, with a contact-cut of 2 λ × 2 λ or larger. The overlapped area must have a margin of at least 1 λ beyond its contact cuts. All available contacts and their design rules are given in Fig. 3.17.

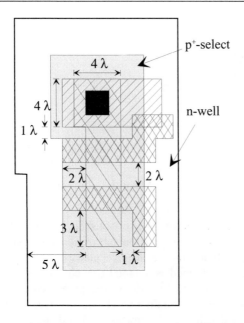

Fig. 3.14 Select area and n-well design rules.

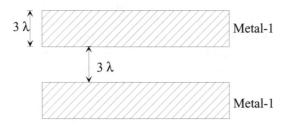

Fig. 3.15 Metal-1 design rules.

A polysilicon-contact has to be separated from a metal-2 contact by at least 1 λ. In contrast to the 3 λ separation rule between two unrelated diffusion regions, a diffusion region must be separated from a diffusion-contact for at least 4 λ. A well-contact can be abutted to the diffusion region in the same well but must be separated from a transistor channel for at least 3 λ. The rules for well contacts also apply to substrate-contacts. All metal-2 contacts (also called vias) must be formed on a "flat" surface, which means that there should not be anything underneath them.

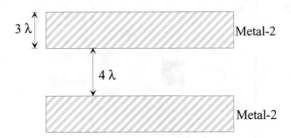

Fig. 3.16 Metal-2 design rules.

Additional design rules have been established to improve the yields of sub-micron fabrication processes. Sub-micron fabrication processes often utilize CMP (Chemical-Mechanical Polishing) to achieve planarity. Effective CMP requires a minimum feature density, defined as the total area covered by the features on a layer divided by the overall area, to be met on the polysilicon and metal layers. For example, MOSIS specifies a minimum polysilicon layer density of 15% and a minimum metal layer density of 30% for its 0.35 and smaller processes. These are known as the minimum density rules. Dummy features can be added to a layout to fill empty spaces to satisfy these rules.

The antenna rules prevent the potential damages induced by the charge collected in the fabrication process on exposed polysilicon and metal features connected to a transistor. The accumulated charge may develop potentials sufficiently high to damage the thin oxide. The antenna rules require that the area of a polysilicon or metal structure over field oxide divided by the area of the transistor gate (thin oxide area) to be less than a given ratio. Since only the layer being processed can collect charge, the antenna rules can be satisfied by breaking long polysilicon or metal features into shorter segments joined by connections formed on a different layer.

3.3   Layout Examples

In this section, we demonstrate layout techniques by providing a number of layout examples. Notice that an interactive design rule checker, such as the one available in the layout tool *magic*, is very helpful for learning layout techniques since it provides instant feedback to the designer. The reader should take advantage of a layout editor with this function if one is available.

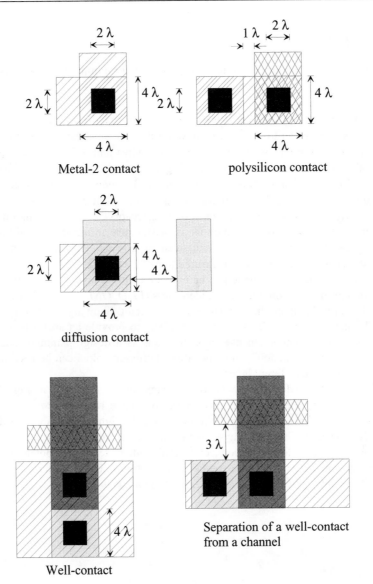

Fig. 3.17 Contact design rules.

Example 3.1

Design a symbolic layout for an inverter $(Z = \overline{A})$ with minimum size transistors and measure its area.

In IC design, the characteristics of a transistor are determined by its channel dimensions. According to the λ-based design rules, the smallest transistor channel is $2\ \lambda$ long (the minimum width of a polysilicon wire) and $3\ \lambda$ wide (the minimum width of a diffusion region). However, in Fig. 3.18, we have increased the width of a transistor to $4\ \lambda$ so that a diffusion-contact ($4\ \lambda \times 4\ \lambda$ required) can be readily made. We often refer to these $2\ \lambda \times 4\ \lambda$ transistors as the minimum size transistors.

An inverter formed of minimum size transistors is called a minimum size inverter. An inverter layout in which the transistors are arranged vertically is shown in Fig. 3.18. Note the use of well and substrate contacts. Its area, defined by the smallest rectangle enclosing it, is measured to be $42\ \lambda \times 15\ \lambda = 630\ \lambda^2$.

A transistor level schematic can be extracted from a layout for verification and simulation purposes. Note that signal identifiers (VDD, GND, A, and Z) are provided on the layout to facilitate the simulation of this circuit. Showing along in Fig. 3.18 is a "stick diagram" of this layout, which is a way to provide an abstract view of the layout. Layout designers can use stick diagrams as a design-planning tool or as a means of "paper and pencil" communication. Different colors can be used in a stick diagram to indicate different layers.

Two slightly different layouts for the same inverter created in Fig. 3.18 are shown in Fig. 3.19. Both of them have an area of $40\ \lambda \times 18\ \lambda = 720\ \lambda^2$. Note that the well/substrate-contacts in the symbolic layout on the right do not abut the diffusion area. The designer should always ensure that in the final composite layout each well/substrate has at least one connection to the adequate voltage potential (V_{DD} for n-type well/substrate and V_{SS} for p-type well/substrate). This is especially important if masks are automatically generated from a symbolic layout.

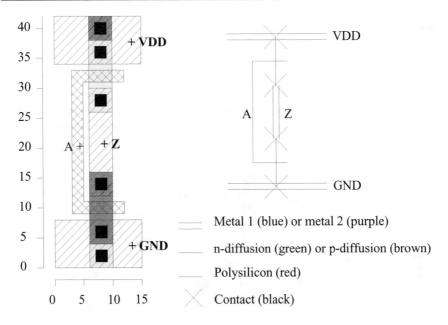

Fig. 3.18 Inverter symbolic layout and its stick diagram.

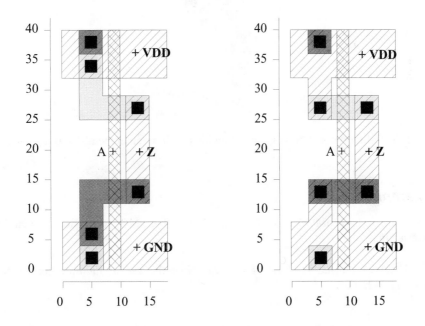

Fig. 3.19 Alternative symbolic layouts for a minimum size inverter.

Example 3.2

Modify the inverter symbolic layout shown in Example 3.1 to allow two metal-1 lines to go through it for routing purposes.

The solution layout is shown in Fig. 3.20. As required, two metal-1 lines go through the layout horizontally. The top one goes through the source of the pMOS transistor and the bottom one passes through the source of the nMOS transistor.

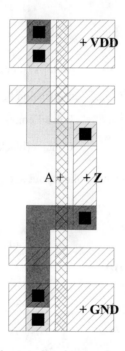

Fig. 3.20 Inverter symbolic layout allowing two horizontal metal-1 routing paths.

Example 3.3

Use a 2 λ × 16 λ pMOS transistor and a 2 λ × 8 λ nMOS transistor to create an inverter symbolic layout.

The layout result is shown Fig. 3.21. Note the use of multiple contact-cuts instead of a big one.

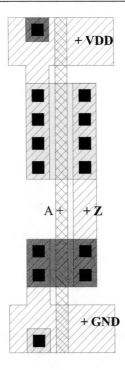

Fig. 3.21 Inverter symbolic layout with wider transistors.

Example 3.4

Create a symbolic layout for a two-input NAND gate $F = \overline{AB}$. Use $2\ \lambda \times 4\ \lambda$ transistors.

Fig. 3.22 shows a layout of the NAND circuit, which is created, without much planning, according to the transistor arrangement in its schematic.

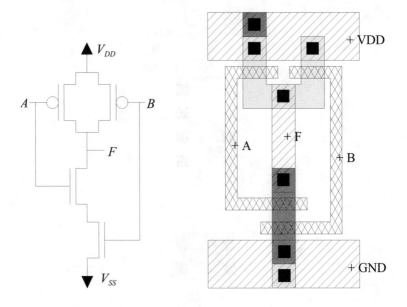

Fig. 3.22 Symbolic layout of a two-input NAND gate.

Example 3.5

Minimize the area of the two-input NAND gate given in Example 3.4 by rearranging the transistors.

Fig. 3.23 shows an alternative layout of the same two-input NAND circuit. Note that this layout arranges the pMOS transistors in a row and places it in parallel to a row of nMOS transistors. This technique, commonly used in standard-cell layouts, simplifies the routing of signal lines and allows the polysilicon wires forming the transistor gates to be readily extended to both the top and bottom of the layout. The area of this layout is about 12% smaller than the one in Fig. 3.22.

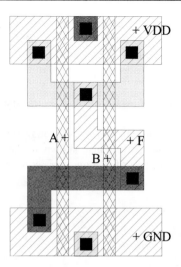

Fig. 3.23 An optimized symbolic layout of a 2-input NAND gate.

An input signal typically drives at least two transistors (one nMOS and one pMOS) in a complementary CMOS circuit. A systematic approach is explained here to arrange the transistors in a layout so that the corresponding pMOS and nMOS transistors driven by the same signal are virtually facing each other across the n and p diffusion spacing.

An example of this layout style has been shown in Fig. 3.23. This arrangement greatly simplifies the routing of signal lines. It is especially attractive in the design of standard-cells when their I/O pins are to be accessed from the top and bottom of the cell (see Chapter 1). The important point in this approach is to arrange the transistors in one of the switch networks (e.g., the pull-up network) as a horizontal sequence. The transistors in the other switch network are then arranged to match the sequence. We use the following example to illustrate this approach.

Example 3.6

Design a symbolic layout for a complementary CMOS circuit that implements $F = \overline{A + BC}$.

The transistor schematic for this function is given in Fig. 3.24. The design objective in this example is to form two rows of transistors. All pMOS transistors should be in one row and all nMOS transistors should be in another row. Furthermore, the pMOS and nMOS transistors should be arranged in the same sequence in their respective rows. We will see that this layout structure eliminates the need to cross input signals.

The transistors in the pull-up and pull-down networks are traced to create two paths, respectively. Without loss of generality, we assume that the first path is created in the pull-up network. The objective of creating this path in the pull-up network is to walk through, without repeating, all pMOS transistors in the pull-up network (Fig. 3.24). Such a path is shown on the pull-up network to consist of the sequence:

transistor $A \rightarrow$ transistor $B \rightarrow$ transistor C.

We must now search for a path that goes through the pull-down network in the same transistor sequence. This path is also shown in the schematic diagram. These two paths determine the sequence of placing transistors in the rows as illustrated in the layout.

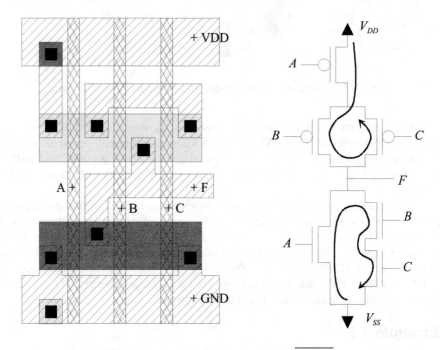

Fig. 3.24 A symbolic layout of $F = \overline{A + BC}$.

In general, a continuous path that goes through all transistors in a switch network may not exist. In this case "breaks" can be inserted into the path. We demonstrate the use of breaks in the following example.

Example 3.7

Design the symbolic layout for the function $F = \overline{A + BD + C}$ with $2\,\lambda \times 8\,\lambda$ transistors.

Refer to the transistor schematic diagram of this logic circuit in Fig. 3.25. No continuous path can go through all the pull-up transistors. A break, indicated by a dotted line, must be used to set up a path consisting of

transistor $A \rightarrow$ transistor $B \rightarrow$ transistor $C \rightarrow$ break $\rightarrow$ transistor D.

In the pull-down network, a matching path is found to be

transistor $A \rightarrow$ transistor $B \rightarrow$ break $\rightarrow$ transistor $C \rightarrow$ transistor D.

The symbolic layout of Fig. 3.25 follows these two sequences to arrange the transistors.

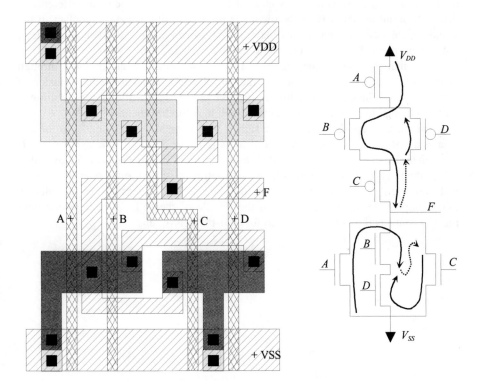

Fig. 3.25 A symbolic layout of $F = \overline{A + BD + C}$.

This approach of arranging transistors into a row can be extended and applied to a circuit in which more than two transistors are driven by the same signal. In such a case, multiple transistor rows can be stacked to maintain the simplicity of input signal routing (see Problem 3.11).

Standard-cells require their I/O pins to be accessible for routing purposes. So far in our example designs, polysilicon lines are used for input purposes. Metal-2 has a better conductivity than polysilicon and is commonly used in a channel/switchbox routing (see Chapter 1). In the following examples we demonstrate a few possible layout styles that provide metal-2 I/O pins.

Example 3.8

Redesign the function given in Example 3.7 into a standard-cell style layout with all I/O pins accessible using metal-2. Double the transistor channel widths from the original 8 λ in Example 3.7 into 16 λ.

Since a standard-cell is designed as a building block to be used in future designs, normally wider transistors are used to provide a larger driving capability. Comparing the layout in Fig. 3.26 with the layout created in Example 3.7, the main difference is that each input signal wire (polysilicon) is provided with a metal-2 contact in the middle of the cell. Since a contact between polysilicon and metal-2 is unavailable, a polysilicon-metal-1 contact is created as an intermediate step.

Note that the layout illustrates how input D and output F are brought, through metal-2 lines, to the top and bottom of the cell. We also demonstrate the inclusion of a metal-2 feed-through (the leftmost metal-2 wire), which allows a path from a routing channel to pass through a standard-cell into another channel.

The use of metal-2 wires to route signals to the contacts in the middle of the cell creates a limitation. It blocks the use of metal-2 in the design of the layout itself. If more than 2 metal layers are available, a few of them can be reserved for over-the-cell routing. This approach allows inter-cell routing to be performed over the cells so that pure routing area, which does not contribute to circuit density, can be reduced. For example, the metal-2 contacts in the middle of the cell can be replaced with metal 3 contacts if metal 3 is reserved for over-the-cell routing.

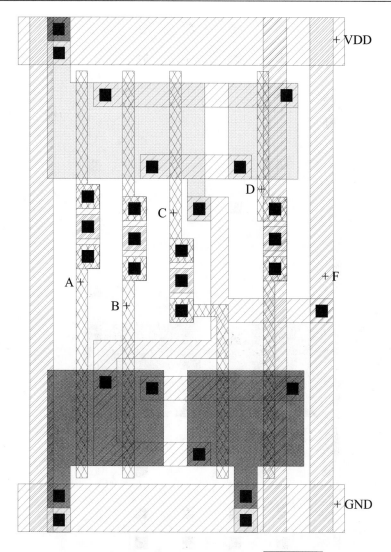

Fig. 3.26 Standard-cell design of $F = \overline{A + BD + C}$.

An alternative layout for this standard-cell is given in Fig. 3.27, in which the metal-2 I/O contacts are made at the top and bottom of the cell. This design style should be used when metal layers are limited. Since routing is done at the borders of the cell, all metal layers can be used within the cell. Avoiding over-the-cell routing has the advantage of preventing unexpected cross coupling between the cell and the overhead routing paths.

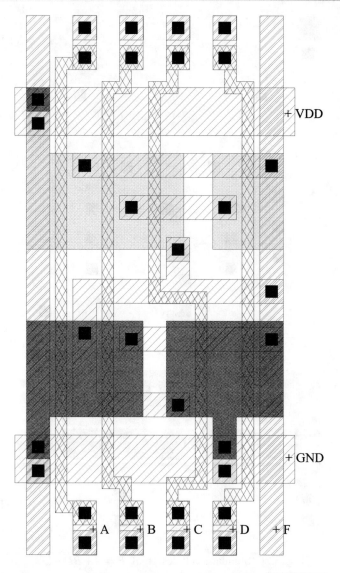

Fig. 3.27 Alternative standard-cell layout of $F = \overline{A + BD + C}$.

The format a layout is described in a computer file depends on the CAD system used. When a design is submitted to a silicon foundry for fabrication, the layout is usually converted into either a Calma GDSII file or Caltech Intermediate Form (CIF) file. GDSII is a binary format and CIF uses a plain ASCII format. References are provided at the end of this chapter for information on these two layout specifications.

3.4   Summary

A simplified manufacturing process of CMOS circuits is presented in this chapter. In order to allow nMOS and pMOS transistors to coexist on the same substrate, portions of the substrate are converted into a type different from the wafer. There are two major operations in integrated circuit manufacturing. In one operation, the properties of the substrate are selectively and locally modified. Another operation selectively deposits materials on the surface of the substrate. An integrated circuit is built by repeating these two operations with different masks created according to a layout designed for the circuit.

While the advances of CAD tools have allowed VLSI designers to design at behavioral or structural levels, full-custom hand-crafted layouts are needed for the creation of library cells and performance critical circuits. Design rules represent the manufacturing limits by a set of width rules and spacing rules. In this chapter, we described a subset of basic scalable λ-based CMOS design rules. A symbolic layout is often used in a design process since it simplifies the layout process and can be translated into a composite layout for which masks are created accordingly. A number of examples are presented to illustrate the use of these design rules to create symbolic layouts. A systematic approach to arrange transistors for improved layout density is explained.

3.5   To Probe Further

VLSI fabrication:

- S. M. Sze, ed., *VLSI Technology*, *2nd Edition*, McGraw-Hill, 1988.

- S. A. Campbell, *The Science and Engineering of Microelectronic Fabrication*, Oxford University Press, 1996.

Scalable λ-based design rules and layout tools:

- C. A. Mead and L. A. Conway, *Introduction to VLSI Systems*. Addison-Wesley, 1980.

- J. K. Ousterhout, G. T. Hamachi, R. N. Mayo, W. S. Scott, and G. S. Taylor, "Magic: A VLSI layout system," *Proceedings of 21st Design Automation Conference*, 1984, pp. 152-159.

- J. P. Uyemura, *Physical Design of CMOS Integrated Circuits Using L-EDIT*, PWS, 1995.

- MOSIS λ-based design rules can be downloaded from http://www.mosis.org

- Information about the popular full-custom layout tool *magic* can be found at http://www.research.digital.com/wrl/projects/magic/magic.html

Prototype fabrication services:

- MOSIS provides prototyping to academic and commercial institutes. Visit http://www.mosis.org for more information.

Commercial CAD tools:

- http://www.cadence.com (Cadence).

- http://www.mentor.com (Mentor Graphics).

- http://www.synopsys.com (Synopsys).

- http://www.avanticorp.com (Avant!).

GDSII and CIF layer specifications:

- C. A. Mead and L. A. Conway, *Introduction to VLSI Systems*, Addison-Wesley, 1980.

- Cadence Design Systems, Inc./Calma. *GDSII Stream Format Manual*, Feb. 1987, Release 6.0, Documentation No. B97E060.

3.6   Problems

For layout design problems, use a layout editor if one is available.

3.1   List three uses of SiO_2 in a CMOS fabrication process.

3.2   Use the SCMOS design rules to draw a symbolic layout for a minimum size transistor. Your answer should include the dimensions of its gate, source, and drain.

3.3   Draw the cross sectional view of the layout shown in Fig. 3.28 along the cutline *K-K'*. Follow the style used in Fig. 3.4 to show your solution.

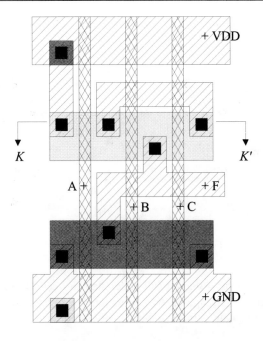

Fig. 3.28 Layout for Problem 3.3.

3.4 Identify the design errors in the layout of Fig. 3.29. The dimensions are given in λ's.

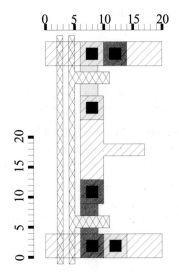

Fig. 3.29 Layout for Problem 3.4.

3.5   Identify the logic function(s) of the layout in Fig. 3.30.

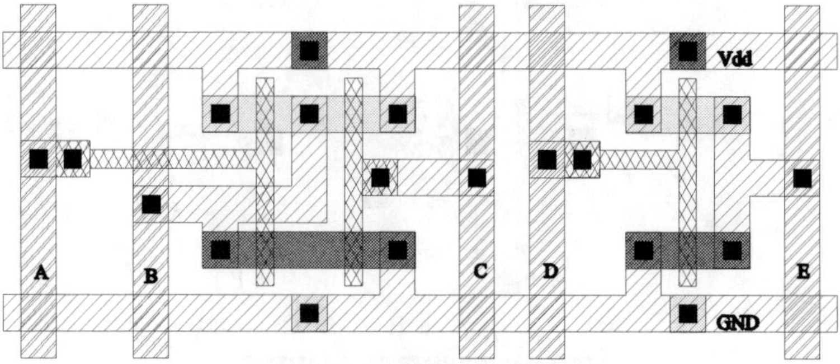

Fig. 3.30 Layout for Problem 3.5.

3.6   Design a layout for a one-bit full-adder standard-cell. The height of this cell
       should be 100 λ. Arrange the transistors into rows so that there is no crossing-
       over of input signals. Minimize the area of your design.

3.7   Explain the importance of applying the polysilicon mask before the diffusion
       mask in a typical CMOS fabrication process.

3.8   Create a CMOS cell for the logic function $Z = \overline{A(B+C)D}$. Use pMOS
       transistors with L = 2 λ, W = 10 λ and nMOS transistors with L = 2 λ, W = 4
       λ. Minimize the area of your design.

3.9   Design a standard-cell layout for a 2-to-1 multiplexer. The height of this cell
       should be 100 λ.

3.10   Design a standard-cell layout for a 2-input XOR. The height of this cell
       should be 100 λ. Minimize the area of your design.

3.11   Extend the transistor arrangement approach to circuits that have more than
       two transistors driven by the same signal so that multiple transistor rows are
       stacked to maintain the simplicity of input signal routing.

3.12   List the order of layer arrangement in a one-polysilicon, five-metal CMOS
       process. Notice that vias are used to connect between metal-1 and metal-2,
       metal-2 and metal-3, metal-3 and metal-4, and metal-4 and metal-5.

Chapter 4 CMOS Circuit Characterization

To design a circuit that works is easy; to design one that works well is hard ...

An integrated circuit is commonly evaluated by its speed, power dissipation, and size, not necessarily in this order. In Chapter 1 we introduced the use of transistor count to estimate the size of an IC. Chapter 3 explained the determination of chip area by measuring its layout. This chapter discusses the evaluation of propagation delay and power dissipation. Since a function often has more than one way of implementation, these criteria can be used, either individually or in combination, to select a structure that suits a specific application.

We begin by reviewing MOSFET behaviors beyond what the simple switch model introduced in Chapter 2 depicts. The use of a circuit-level simulator (SPICE) to simulate and evaluate a CMOS circuit is briefly discussed. Following the estimation of CMOS circuit parasitic resistance and capacitance values, we develop a MOSFET resistive switch model. The use of this model in propagation delay estimation is described.

Not all parasitic parameters are available until the physical layout of the design is available. We present a simplified delay model called the τ-model which can be used to explore propagation delay issues without getting involved with the physical layout details. The use of this model to design optimized buffers for driving large loads is demonstrated. The power dissipation evaluation of a CMOS circuit is explained. Finally a potential problem in CMOS circuits called latch-up and its prevention are discussed.

4.1   MOSFET Theory

The simple MOSFET switch model has allowed CMOS logic circuits to be designed and analyzed easily. However, an enhanced model that incorporates some non-idealistic MOSFET behaviors is needed in the estimation of delay time and power dissipation. The following nMOS circuit, originally presented in Fig. 2.2 for the introduction of the simple switch model, is reproduced in Fig. 4.1 for a more in-depth analysis of the nMOS transistor.

Instead of restricting the operating voltages of the circuit in Fig. 4.1 to only V_{SS} and V_{DD}, as we have previously done in the development of the simple switch model, this circuit will now be analyzed with both v_I and v_D varying continuously and independently between V_{DD} and V_{SS}. A family of curves showing the i_D-v_{DS} characteristics of the nMOS transistor measured at different v_I's are given in Fig. 4.2. Notice that $v_I = v_{GS}$, the gate to source voltage of the nMOS transistor.

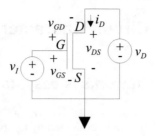

Fig. 4.1 Setup for measuring the i_D-v_{DS} characteristics of an nMOS transistor.

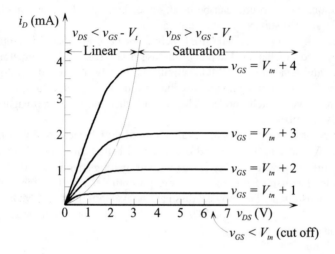

Fig. 4.2 The i_D-v_{DS} characteristics for the nMOS transistor.

Fig. 4.2 shows the three operation regions of the nMOS transistor, which are described as follows.

The cutoff region

The nMOS transistor is turned off, as correctly depicted by the simple switch model, when $v_I = v_{GS} \leq V_{tn}$, the threshold voltage of the transistor. In this cutoff region, $i_D \approx 0$.

Now assume that the transistor is in its cutoff region and v_I is gradually increased. When v_I is increased to above V_{tn}, a channel is induced under the gate and the transistor turns on with $i_D > 0$. The operation of an nMOS transistor when it is turned on, however, is further divided into two distinct regions.

The linear region

When the transistor is turned on and v_{DS} is kept small enough, the channel is continuous from the source to the drain. In order to maintain a channel continuously between the source and drain ends, v_{GS} and v_{GD} must be kept above V_{tn}. Since $v_{GD} = v_{GS} - v_{DS}$, the condition of $v_{GD} > V_{tn}$ is transformed into $v_{GS} - v_{DS} > V_{tn}$. This is the familiar condition for linear region operation,

$$v_{DS} < v_{GS} - V_{tn}. \tag{4.1}$$

In the linear region, the i_D-v_{DS} characteristics can be described by the following equation:

$$i_D = k_n \frac{W}{L} \left[(v_{GS} - V_{tn})v_{DS} - \frac{1}{2}v_{DS}^2 \right], \tag{4.2}$$

where $k_n = \dfrac{\mu_n \varepsilon_{ox}}{t_{ox}}$ is a process dependent parameter called the transconductance parameter, μ_n is the electron mobility (1350 cm²/V-s), ε_{ox} is the gate oxide permittivity (3.5×10^{-13} F/cm), t_{ox} is the thickness of the gate oxide (e.g., 0.02 μm), W is the channel width, and L is the channel length.[1] If v_{DS} is sufficiently small, the v_{DS}^2 term can be ignored and i_D is linearly dependent on v_{DS}. This operation region is thus referred to as the linear region.

The saturation region

For a fixed v_{GS}, if v_{DS} is increased to exceed $(v_{GS} - V_{tn})$, the voltage difference between the gate and the drain (V_{GD}) drops below V_{tn} and the channel depth at the drain end decreases to almost zero. Under this condition, the channel is said to be pinched off and the current through the channel becomes independent of v_{DS} and essentially remains a constant. In other words, the channel current saturates. In the saturation region, the i_D-v_{DS} characteristics can be described by the following equation:

$$i_D = \frac{1}{2}k_n \frac{W}{L}(v_{GS} - V_{tn})^2 \tag{4.3}$$

which is independent of v_{DS}.

In reality, this equation is only an approximation since as v_{DS} increases beyond $(v_{GS} - V_{tn})$, the channel pinch-off point moves slightly away from the drain and the effective channel length is reduced. This effect, known as channel-length modulation,

---

[1] The mobility is temperature dependent.

results in a higher i_D as v_{DS} increases since the current is inversely proportional to the channel length.

We can apply the duality principle between nMOS and pMOS transistors to derive the operation regions of pMOS transistors. These operation regions are presented along with their nMOS counterparts in Fig. 4.4. The circuit that is used to determine the pMOS i_D-v_{DS} characteristics is shown in Fig. 4.3. For the convenience of analyzing CMOS circuits, i_D assumes a direction from source to drain for the pMOS transistor, which is different from Fig. 4.1 (from drain to source).

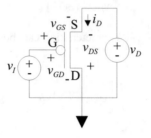

Fig. 4.3 Setup for measuring the i_D-v_{DS} characteristics of a pMOS transistor.

Operating Regions	nMOS ($v_{GS} > 0$, $v_{DS} > 0$, $V_{tn} > 0$)	pMOS ($v_{GS} < 0$, $v_{DS} < 0$, $V_{tp} < 0$)
Cutoff	$v_{GS} \leq V_{tn}$, $i_D = 0$	$v_{GS} \geq V_{tp}$, $i_D = 0$
Linear	$v_{GS} > V_{tn}$ $v_{DS} < v_{GS} - V_{tn}$ $i_D =$ $k_n \dfrac{W}{L}\left[(v_{GS} - V_{tn})v_{DS} - \dfrac{1}{2}v_{DS}^2\right]$	$v_{GS} < V_{tp}$ $v_{DS} > v_{GS} - V_{tp}$ $i_D =$ $k_p \dfrac{W}{L}\left[(v_{GS} - V_{tp})v_{DS} - \dfrac{1}{2}v_{DS}^2\right]$
Saturation	$v_{GS} > V_{tn}$ $v_{DS} > v_{GS} - V_{tn}$ $i_D = \dfrac{1}{2}k_n \dfrac{W}{L}(v_{GS} - V_{tn})^2$	$v_{GS} < V_{tp}$ $v_{DS} < v_{GS} - V_{tp}$ $i_D = \dfrac{1}{2}k_p \dfrac{W}{L}(v_{GS} - V_{tp})^2$

Fig. 4.4 Operation regions of nMOS and pMOS transistors.

In Fig. 4.4, $k_p = \dfrac{\mu_p \varepsilon_{ox}}{t_{ox}}$ is the transconductance of the pMOS transistor. $\mu_p = 480$ cm^2/Vs and is the mobility of holes. In general, the transistors switched on in a steady state CMOS circuit operate in their linear regions. An nMOS transistor, operating in its linear region, has a channel resistance of

$$r_{DSn} = \left| \frac{\partial v_{DSn}}{\partial i_D} \right| = \left[k_n \frac{W_n}{L_n} (v_{GSn} - V_{tn}) \right]^{-1} \quad \text{when } v_{DSn} = 0 \qquad (4.4)$$

Similarly, the pMOS transistor has a linear region channel resistance of

$$r_{DSp} = \left| \frac{\partial v_{DSp}}{\partial i_{Dp}} \right| = \left[k_p \frac{W_p}{L_p} (v_{GSp} - V_{tp}) \right]^{-1} \quad \text{when } v_{DSp} = 0 \qquad (4.5)$$

Typically, $\mu_p = 0.4 \ \mu_n$, and thus the linear channel resistance of a pMOS transistor is at least 2.5 times higher than an nMOS transistor with the same channel length (L) and width (W). Notice the use of a subscript (n or p) in the above equations to distinguish between nMOS and pMOS transistors.

4.2   Voltage Transfer Characteristic

While digital systems are designed to deal with only binary signals, practical signals cannot jump from one level to another instantaneously. Instead, signals take a finite time to change and experience propagation delays. It is thus important to develop an understanding of what is going on in the CMOS circuit when signals change before we develop an improved transistor switch model to evaluate delay times.

Voltage transfer characteristic describes the voltage relationship between the input and output signals of a circuit. We will use an inverter as a general CMOS complementary structure to derive the voltage transfer characteristic of CMOS complementary logic circuits. Notice that this discussion goes beyond the analysis of an inverter since all complementary logic circuits operate in a manner similar to the inverter.

Fig. 4.5 shows the inverter that we are going to analyze and its voltage transfer function, which is divided into 5 regions (a to e). First recall that the simple switch model is valid when a transistor is turned off so regions a and e can be readily determined.

In region a, the nMOS transistor is off. During steady state, there is no current flowing through the inverter so $v_Z = V_{DD}$. Therefore, $v_{DSp} = 0 > v_{GSp} - V_{tp}$, and the pMOS transistor is in its linear operating mode. Symmetrically, in region e, the pMOS transistor is off, $v_Z = 0$ and the nMOS transistor is in its linear operating mode.

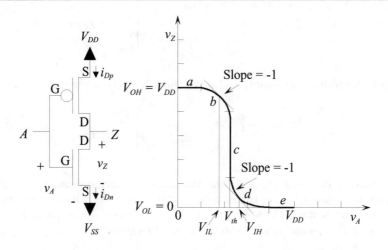

Fig. 4.5 CMOS inverter and its voltage transfer characteristic.

In regions b, c, and d, both nMOS and pMOS transistors are turned on and the inverter is in transition. In order to help visualize the relationship between the transistors, we superimposed the i_D-v_{DS} curve sets of the pMOS and nMOS transistors in Fig. 4.6 by recognizing the following relations.

$$i_{Dp} = i_{Dn} \tag{4.6}$$

$$v_A = v_{GSn}, \; v_Z = v_{DSn} \tag{4.7}$$

$$v_A - V_{DD} = v_{GSp}, \; v_Z - V_{DD} = v_{DSp} \tag{4.8}$$

We provide a qualitative analysis of the voltage transfer characteristic with respect to Fig. 4.6 before its quantitative development. In order to satisfy (4.6), for a given v_A, the circuit must operate at the intersecting point between the nMOS and pMOS curves corresponding to v_A. If v_A increases beyond V_{tn}, the nMOS transistor begins to turn on and forms a voltage divider with the pMOS transistor, causing output v_Z to decrease.

Graphically we can see that the nMOS transistor is in its saturation region and the pMOS transistor is in its linear region. This corresponds to region b. As v_A continues to increase, v_Z continues to decrease and the circuit moves toward the center of the graph (region c) in which both transistors are in saturation. Further increasing the input voltage moves the circuit into region d in which the nMOS transistor is in its linear mode and the pMOS transistor is in saturation. Finally, the circuit settles into region e when v_a goes beyond $V_{DD} + V_{tp}$ which turns off the pMOS transistor.

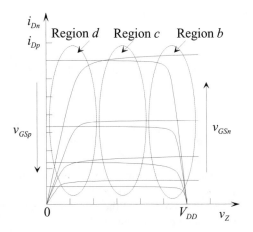

Fig. 4.6 Relationship between the nMOS and pMOS transistors in an inverter.

We now derive the boundary points that separate regions and the expression of v_Z as a function of v_A.

Region a:

The simplified transistor switch model is valid in this region. The nMOS transistor is off and the pMOS transistor is in its linear region with $v_{DSp} = 0$.

$$v_A \le V_{tn}; \quad v_Z = V_{DD}; \quad i_{Dp} = i_{Dn} = 0 \tag{4.9}$$

Region b:

The nMOS transistor is turned on so $v_A \ge V_{tn}$. This establishes the lower bound of v_A in region b. The pMOS transistor is in its linear region, which implies $v_{DSp} \ge v_{GSp} - V_{tp}$. This can be rewritten in terms of v_A and v_Z as $v_Z - V_{DD} \ge v_A - V_{DD} - V_{tp}$.

The upper bound of v_A in this region is thus established as $(v_Z + V_{tp})$. The relation of v_A and v_Z is given by the following current equation:

$$\begin{aligned}
i_{Dp} &= k_p \frac{W_p}{L_p} \left[(v_A - V_{DD} - V_{tp})(v_Z - V_{DD}) - \frac{1}{2}(v_Z - V_{DD})^2 \right] \\
&= \frac{1}{2} k_n \frac{W_n}{L_n} (v_A - V_{tn})^2 = i_{Dn}
\end{aligned} \tag{4.10}$$

Substituting v_A with $(v_Z + V_{tp})$ in the above current equation determines the output voltage when the inverter is leaving region b and going into region c. Alternatively, the boundary between regions b and c can also be found by considering region c, as we are going to demonstrate next.

Region c:

In this region both transistors are in saturation. Equating the current expressions we have

$$i_{Dp} = \frac{1}{2}k_p \frac{W_p}{L_p}(v_A - V_{DD} - V_{tp})^2 = \frac{1}{2}k_n \frac{W_n}{L_n}(v_A - V_{tn})^2 = i_{Dn} \qquad (4.11)$$

The above equation has only one variable v_A, which implies that region c occurs at a single value of v_A. Due to the continuity of the voltage transfer characteristic, this value, defined as the logic threshold V_{th} of the inverter, is also the upper bound of v_A in region b and the lower bound of v_A in region d.

Region d:

The pMOS transistor is turned on so $v_{GSp} = v_A - V_{DD} \le V_{tp}$ or $v_A \le V_{DD} + V_{tp}$. The nMOS transistor is in its linear region which implies $v_{DSn} \le v_{GSn} - V_{tn}$, which can be rewritten in terms of v_A and v_Z as $v_Z \le v_A - V_{tn}$.

The lower bound of v_A in this region is thus established as $(v_Z + V_{tn})$. The relation of v_A and v_Z is given by the following current equation:

$$i_{Dn} = k_n \frac{W_n}{L_n}\left[(v_A - V_{tn})v_z - \frac{1}{2}v_z^2\right] = \frac{1}{2}k_p \frac{W_p}{L_p}(v_A - V_{DD} - V_{tp})^2 = i_{Dp} \qquad (4.12)$$

Substituting v_A with $(v_Z + V_{tn})$ in the current equation determines the output voltage when the inverter is leaving region d into region c. Alternatively, the boundary between regions c and d can also be found by considering region c, as we have shown previously in region c.

Region e:

Again, the simplified transistor switch model is valid in this region. The nMOS transistor is in its linear region with $v_{DSn} = 0$. The pMOS transistor is off.

$$v_A \ge V_{DD} + V_{tp}; \; v_Z = 0; \; i_{Dp} = i_{Dn} = 0 \qquad (4.13)$$

In Fig. 4.5 the input voltage at which both transistors are in saturation (region b) is defined as the logic threshold V_{th}. Ideally the logic threshold V_{th} should be set at the mid point of the logic swing or $V_{DD}/2$. This occurs only when

$$k_n \frac{W_n}{L_n} = k_p \frac{W_p}{L_p} \tag{4.14}$$

in which case the nMOS and pMOS transistors are said to be matched. If k_p is 0.4 times the value of k_n, the width of the pMOS transistor must be made 2.5 times that of the nMOS transistor in order to satisfy the matching condition, assuming both have the same channel length.

Two other points on the voltage transfer curve determine the noise margins of the inverter. These are V_{IH}, which is the lowest input voltage level considered to be a logic 1, and V_{IL}, which is the highest input voltage level considered to be a logic 0. V_{IH} and V_{IL} are defined as the two points on the transfer curve with unity incremental gain (slope = -1 V/V).

The noise margin for high input, NM_H, and the noise margin for low input, NM_L, can now be determined as:

$$NM_H = V_{OH} - V_{IH}; \; NM_L = V_{OL} - V_{IL} \tag{4.15}$$

4.3   Circuit-Level Simulation

The use of simulation tools to verify and evaluate the operation of an IC is an essential step in the design process. On the one hand it verifies that the design will meet specifications. On the other hand a simulation provides additional insight into the operation so that it can be fine tuned prior to its fabrication.

Due to the complexity of VLSI circuits, switch-level simulation which considers transistors as switches is often used. In many cases, switch level simulation is still inefficient and gate level or block level simulation has to be used. However, circuit-level simulation is an indispensable tool to more accurately determine circuit behaviors. Among the various circuit-level simulation programs, SPICE is by far the most popular one. SPICE (Simulation Program with Integrated Circuit Emphasis) was developed at the University of California, Berkeley, in the early 1970's. It has been developed into a number of free or low cost and commercial products.

It is not the objective here to explain to the reader how to use SPICE. There are many good books on this subject and a few of them are listed at the end of this chapter. We intend to describe the models that SPICE use for the MOSFET and to illustrate the use of SPICE in the design of a CMOS logic circuit.

SPICE uses rather complicated MOSFET models in the simulation of CMOS circuits. There are two issues in a SPICE device model. The first is the structure of the model developed for a category of devices. The simulator used must support the model structure to be used. Second, the values of a model specifically determined for a fabrication technology must be obtained.

Semiconductor manufacturers spend a significant amount of effort to model their products. MOSFET models are often the proprietary property of their developers. In order to facilitate the exploration of CMOS circuit characteristics using SPICE, example values of two MOSFET models, one for nMOS transistors and one for pMOS transistors, are provided below. Special MOSFET models have been developed for deep sub-micron designs.[2] The reader is reminded that circuit simulation can only be as accurate as its models. Be sure to acquire and use the models supplied by your selected IC manufacturer.

```
.MODEL nfet NMOS LEVEL=2 LD=0.12U TOX=200E-10
+NSUB=1.0E+16 VTO=0.8 KP=5.0E-5 GAMMA=0.45 PHI=0.6
+UO=700 VMAX=70000 XJ=0.6U LAMBDA=.02 NSS=0.0
+TPG=1.00 RSH=0 CGSO=0.9E-10 CGDO=0.9E-10 CJ=1.4E-4
+MJ=0.6 CJSW=6.5E-10 MJSW=0.3

.MODEL pfet PMOS LEVEL=2 LD=0.18U TOX=200E-10
+NSUB=1.5E+15 VTO=-0.7 KP=2.0E-5 GAMMA=0.45 PHI=0.6
+UO=250 VMAX=34600 XJ=0.4U LAMBDA=.04 NSS=0.0
+TPG=-1.00 RSH=0 CGSO=1.4E-10 CGDO=1.4E-10 CJ=2.4E-4
+MJ=0.5 CJSW=3.5E-10 MJSW=0.3
```

The description of a circuit to be simulated is called a SPICE deck.[3] Each deck should include the following parts: a list of the components and how they are connected; component model parameters; power source; input signals; and control cards instructing SPICE to perform simulation. A SPICE deck can be generated by extracting the circuit and parasitic components from a layout. The parasitic extraction of deep sub-micron (DSW) ICs is currently a major research problem.

We present the SPICE deck of an inverter, the schematic of which is shown in Fig. 4.7, as an example of performing SPICE simulation. Each circuit node is labeled with a unique number. V_{DD} and V_{SS} are conventionally labeled as 1 and 0, respectively. Note that each transistor has its channel dimensions specified.

---

[2] See http://www-device.eecs.berkeley.edu/~bsim3/

[3] In the 70s, circuit descriptions were put on a deck of punch cards. The cards became obsolete but the name stayed.

```
* SPICE DECK created for a CMOS inverter Z=not(A)

M1 1 2 3 1 pfet L=1.0U W=4.0U
M2 3 2 0 0 nfet L=1.0U W=4.0U
C3 3 0 0.028000PF

*    A    2
*    GND  0
*    VDD  1
*    Z    3

* CMOS SPICE model parameters

.MODEL nfet NMOS LEVEL=2 LD=0.12U TOX=200E-10
+NSUB=1.0E+16 VTO=0.8 KP=5.0E-5 GAMMA=0.45 PHI=0.6
+UO=700 VMAX=70000 XJ=0.6U LAMBDA=.02 NSS=0.0
+TPG=1.00 RSH=0 CGSO=0.9E-10 CGDO=0.9E-10 CJ=1.4E-4
+MJ=0.6 CJSW=6.5E-10 MJSW=0.3

.MODEL pfet PMOS LEVEL=2 LD=0.18U TOX=200E-10
+NSUB=1.5E+15 VTO=-0.7 KP=2.0E-5 GAMMA=0.45 PHI=0.6
+UO=250 VMAX=34600 XJ=0.4U LAMBDA=.04 NSS=0.0
+TPG=-1.00 RSH=0 CGSO=1.4E-10 CGDO=1.4E-10 CJ=2.4E-4
+MJ=0.5 CJSW=3.5E-10 MJSW=0.3

* Power source
vps 1 0 dc 5

* Input waveform
vA 2 0 dc 0 pwl(0ns 0 0.01ns 5 10ns 5
+10.01ns 0 20ns 0 20.01ns 5 30ns 5)

* Perform a transient analysis with a step size
* of .1 ns for the duration of 40 ns.
.tran 0.1ns 40ns

* Provide values and plots for V(A) and V(Z)
.print tran v(2) v(3)
.plot tran v(2) v(3)

.end
```

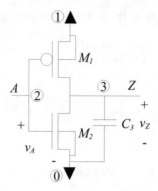

Fig. 4.7 Inverter circuit for SPICE simulation.

The waveforms of both the input ($V(2)$ or $V(A)$) and output ($V(3)$ or $V(Z)$) produced by a SPICE simulation are shown in Fig. 4.8.

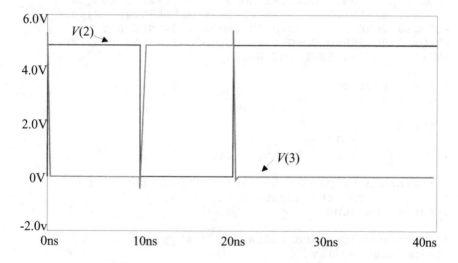

Fig. 4.8 SPICE simulation result of an inverter.

4.4   Improved MOSFET Switch Model

In this section we seek to represent MOSFETs as resistive switches. A transistor switch in the off state has a source-drain resistance which is high enough to be considered as an open circuit in general applications. When the transistor is turned on, we would like to model it as a resistance charging or discharging a capacitance.

Timing models can then be developed to estimate the delays of logic circuits by observing the switch behavior of the transistors. Fig. 4.9 illustrates this resistive switch model applied to an NMOS transistor. The transistor in its on state is modeled by a channel resistance R_n connected in series with a closed switch.

In order to develop an approach to determine the channel resistance of a transistor switch model, we have to look at its dynamic behavior when it is switching. The channel resistance of MOSFETs operating in the linear region were developed and provided in equations (4.4) and (4.5). However, the discussion of the voltage transfer characteristic has shown that a transistor, after it has been switched on, first operates in its saturation region before settling in its linear region.

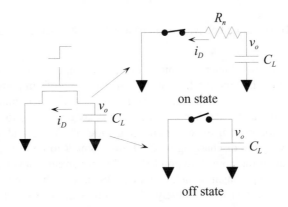

Fig. 4.9 Resistive switch model of an nMOS transistor.

We assume that the transistor gate is driven by an ideal step signal (zero rise time) as shown in Fig. 4.9. Fig. 4.10 describes the dynamic behavior of the nMOS transistor in response to the step signal. Assume that $v_o = V_{DD}$ just prior to the rising edge of the input signal (time = 0–). The transistor is operating at point A (cut off) and $i_D = 0$. Right after the input signal has gone up to V_{DD} (time = 0+), $v_{GS} = V_{DD}$ and the operating point of the transistor is moved up to point B.

The assumption of an ideal step signal implies that this jump is instantaneous. In practice we need to satisfy this assumption by using signals that have reasonably fast rise and fall times. As long as v_{GS} remains at V_{DD}, the operating point of the transistor will have to move along its i_D-v_{DS} curve corresponding to $v_{GS} = V_{DD}$, which is the trajectory shown in Fig. 4.10. Initially the transistor operates in its saturation region until point C is reached. During this stage, the transistor is acting like a constant current source and the capacitance C_L is being discharged. At point C, $v_o = v_{DS} = v_{GS} - V_{tn} = V_{DD} - V_{tn}$, which is the boundary between linear and saturation regions. From this point on, the transistor works in its linear region and acts like a resistance, through which the capacitance C_L continues to be discharged. Eventually, point E is reached, the switching is completed and the discharging current stops.

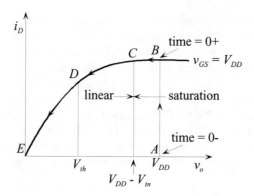

Fig. 4.10 Dynamic behavior of an nMOS transistor being turned on.

Fig. 4.10 shows two important points. First, if the input signal has a reasonably fast rise time so that the time it takes for the transistor to go from point A to point B is insignificant, the i_D - v_{DS} curve corresponding to $v_{GS} = V_{DD}$ determines the current discharging C_L and thus the discharging time. The delay time for v_o to respond to the input change is proportional to the discharging time of C_L. Second, there are two stages involved with the discharging of C_L. From point B to point C, the saturation current is involved. From point C to point E, the linear current is involved.

Let us define the propagation delay of a CMOS circuit. This seemingly easy task is complicated by the fact that a variety of definitions exist, each of which has its merits. Again, an inverter is used as a general CMOS structure to present the following discussion.

Our definition of propagation delay is shown in Fig. 4.11 in which the input and output signals of an inverter ($Z = \overline{A}$) are shown on the same time axis. At t_1 input v_A goes across the logic threshold V_{th} of the inverter. The output v_Z responds to the change at v_A and it crosses the logic threshold V_{th} of the next stage driven by the inverter at t_2. The high-to-low propagation delay (at the output) is defined as $t_{PHL} = t_2 - t_1$. The low-to-high propagation delay (t_{PLH}) is defined similarly.

Unless the pull-up and pull-down switch networks are matched, the logic threshold V_{th} is a function of the transconductances and sizes of the transistors. For the sake of a good noise margin, a reasonable design should not have a logic threshold too far away from the mid point of the logic swing; so $V_{th} = V_{DD}/2$ is often assumed.

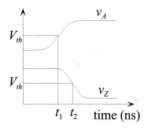

Fig. 4.11 Definition of propagation delay.

To follow up with the discussion in Fig. 4.10, if we assume that the input is an ideal signal, $V_{th} = V_{DD}/2$, and $V_{tn} = 0.2V_{DD}$, the high-to-low propagation delay t_{PHL} can be derived by considering the time it takes to discharge the capacitance C_L from V_{DD} to $V_{DD}/2$:

$$t_{PHL} = \frac{1.6C_L}{k_n(W_n / L_n)V_{DD}} \tag{4.16}$$

Similarly, we can derive the low-to-high propagation delay t_{PLH} of a pMOS transistor charging a capacitance C_L with an initial voltage of 0 as

$$t_{PLH} = \frac{1.6C_L}{k_p(W_p / L_p)V_{DD}} \tag{4.17}$$

The reader is reminded that the usefulness of equations (4.16) and (4.17) is limited to the specific definitions of t_{PHL} and t_{PLH} and various assumptions. According to these equations the propagation delay is determined by the load capacitance C_L. Notice that the propagation delay can be manipulated by changing the circuit parameters W, L, and V_{DD}.

To provide a very simple RC model to estimate delay times, we also define t_{PHL} and t_{PLH} using the resistive switch models shown in Fig. 4.12. In Fig. 4.12, the nMOS transistor that is involved with discharging C_L is replaced by a channel resistance R_n. Similarly, a channel resistance R_p represents a pMOS transistor charging C_L. We define the propagation delays as

$$t_{PHL} = R_n \times C_L \tag{4.18}$$

$$t_{PLH} = R_p \times C_L. \tag{4.19}$$

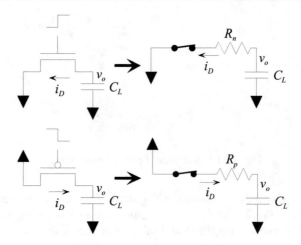

Fig. 4.12 RC timing models.

Comparing these equations with (4.18) and (4.19), respectively, gives the following expressions for R_n and R_p, which are called the effective channel resistance of the transistors:

$$R_n = \frac{1.6}{k_n (W_n / L_n) V_{DD}} = R_{sn} \frac{L_n}{W_n} \tag{4.20}$$

$$R_p = \frac{1.6}{k_p (W_p / L_p) V_{DD}} = R_{sp} \frac{L_p}{W_p} \tag{4.21}$$

R_{sn} and R_{sp} are called the n-channel sheet resistance and p-channel sheet resistance, respectively, which are physical parameters determined by the manufacturing process.

4.5   Resistance/Capacitance Estimation

In the previous section we have derived the use of effective channel resistance and load capacitance in the calculation of propagation delay. Parasitic resistance and capacitance are also very important in the estimation of propagation delays. We will thus introduce the approach to estimate resistance and capacitance to be used in propagation delay analysis.

Resistance

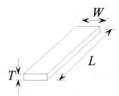

Fig. 4.13 Determination of resistance value.

The resistance of a uniform piece of conducting material shown in Fig. 4.13 is expressed as

$$R = \left(\frac{\rho}{T}\right)\left(\frac{L}{W}\right)\Omega, \tag{4.22}$$

where ρ = resistivity, T = thickness, L = length, and W = width. Recall that VLSI design is 2-dimensional from the designer's viewpoint and thickness is determined by the fabrication process so T may be considered a constant. Also, the resistivity ρ is a physical property of the conductor material and is a constant. The above expression can thus be rewritten as

$$R = R_s\left(\frac{L}{W}\right)\Omega, \tag{4.23}$$

where $R_s = \dfrac{\rho}{T}$ is called the sheet resistance (Ω/square or $\Omega/\square$) of a specific fabrication process. The resistance of a conductor can thus be found by multiplying its sheet resistance R_s and length to width ratio L/W. The sheet resistance values of the materials used in an example 1 µm CMOS processes are given in Fig. 4.14. Notice that Fig. 4.14 also provides the effective sheet resistance values of both n-channel and p-channel transistors, which should not be confused with that of n-diffusion and p-diffusion areas. In general, effective channel sheet resistance increases as V_{DD} is reduced. The reader is again reminded to acquire and use resistance and capacitance values for your selected fabrication process.

Layer	Ω/square
Metal 1 and 2	0.05
n-type Diffusion	2
p-type Diffusion	2
Polysilicon	4
nMOS Channel (effective)	6 K
pMOS Channel (effective)	15 K
Capacitance	**fF/μm^2**
nMOS and pMOS gate	1
Diffusion (area)	0.8
Diffusion (periphery)	0.3 (fF/μm)
Polysilicon	0.1
Metal 1	0.05
Metal 2	0.02

Fig. 4.14 Example resistance and capacitance values.

In a layout design, wires often consist of rectangular segments. The resistance of these wires can be found by evaluating and adding the resistance of their rectangular segments. This is demonstrated in the following example.

Example 4.1

Calculate the resistance of the polysilicon wire shown in Fig. 4.15. This polysilicon wire is to be fabricated with a 1 μm technology.

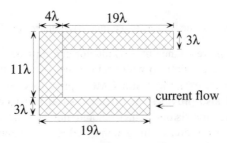

Fig. 4.15 Resistance value calculation of a polysilicon wire.

The polysilicon wire is divided into three rectangular segments (19 $\lambda \times$ 3 λ, 11 λ $\times$ 4 λ, and 19 $\lambda \times$ 3 λ). The total resistance of this wire, with R_s = 4 Ω/sq. for polysilicon from Fig. 4.14, is calculated as

$$R_{poly} = 4\Omega / \text{square} \times \left(\frac{19}{3} + \frac{11}{4} + \frac{19}{3}\right) \text{squares} = 61.67\Omega.$$

Notice that this is a slight overestimation of the wire resistance since the turning corners have reduced resistance.[4] We do not want to burden our estimation with calculating the corner resistance separately due to the fact that they constitute only an insignificant portion of the entire wire length. The estimated resistance of a path also varies slightly depending on how it is broken up into segments.

Capacitance

Another important parasitic element in CMOS circuits is capacitance. Consider the structure in Fig. 4.16, which has a conductor plate of area $(L \times W)$ placed on top of a base conductor plate with a much larger area. These plates are separated by a distance of T. This structure is ubiquitous in CMOS circuits and can be applied to all layers deposited on top of the substrate. For example, the top plate could be a piece of metal or polysilicon while the base plate is the substrate.

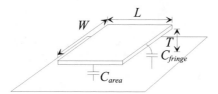

Fig. 4.16 Determination of capacitance value.

The calculation of the top plate capacitance with respect to the base can be divided into two parts. The first part is the capacitance caused by the area of the top plate which can be calculated as

$$C_{area} = \text{Capacitance per unit area} \times L \times W. \qquad (4.24)$$

The second part is the capacitance caused by the edges of the top plate, which is called the fringe capacitance and can be calculated as

$$C_{fringe} = \text{Capacitance per unit length} \times 2(L + W). \qquad (4.25)$$

Both parameters, "capacitance per unit area" and "capacitance per unit length", are dependent on the dielectric constant of the material separating the two plates and the spacing T between the plates. Recall again that the vertical spacing between a layer and the substrate is, from the viewpoint of a designer, a constant. These two

---

[4] A corner square has a sheet resistance of ~0.6 R_s.

parameters can then be determined for each layer in a given manufacturing process. In order to simplify the estimation of parasitic capacitance, we adopt the following general equation

$$C = C_u \times L \times W \qquad (4.26)$$

where C_u is the capacitance per unit area (pF/μm^2 or fF/μm^2) which has been "adjusted" to cover both area and fringe capacitance.

Fig. 4.14 lists the unit area capacitance of each layer with respect to the substrate. Layer to layer capacitance caused by overlapping two layers (e.g., metal 1 and metal 2) is normally small enough to be ignored unless the overlapping is substantial which should be avoided if possible.

Notice that due to the different thicknesses of gate oxide and field oxide, the unit area capacitance of a transistor gate is very different from that of a polysilicon wire.

With the exception of diffusion layers, the capacitance of all layers can be estimated with equation (4.26). In the case of the diffusion layers, the fact that they are *buried* in the substrate and the capacitance is caused by a depletion region located at the junction requires them to be calculated with a different method.

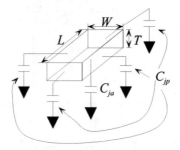

Fig. 4.17 Diffusion capacitance.

Fig. 4.17 shows the relationship of a diffusion slab within the substrate. Both the periphery wall (C_{jp}) and the bottom (C_{ja}) of the slab contribute to the parasitic capacitance. The bottom capacitance is calculated by multiplying its area ($L \times W$) with the unit area capacitance. The periphery capacitance C_{jp} is, on the other hand, calculated by multiplying its linear length ($2(L \times W)$) with the unit length periphery capacitance (pF/μm). This is another example of keeping the design in a 2-dimensional space. The unit length periphery capacitance has the thickness T of the diffusion slab figured in. The overall diffusion capacitance is thus

$$C_d = C_{u(jp)} \times 2(L+W) + C_{u(ja)} \times (L \times W) \qquad (4.27)$$

Notice that the diffusion junction area capacitance assumes a zero bias across the junction so that the depletion region (acting as a dielectric layer) is at its minimum width to ensure a worst case estimation.

Example 4.2

Estimate the resistance and capacitance values for the symbolic layout of a two-input NAND gate shown in Fig. 4.18. Assume a 1 μm technology with the values in Fig. 4.14 is used.

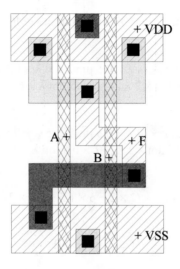

Fig. 4.18 Resistance and capacitance value estimation.

Effective channel resistance:

nMOS: $R_n = 6 \text{ K}\Omega \times 2 \, \lambda/4 \, \lambda = 3 \text{ K}\Omega$

pMOS: $R_p = 15 \text{ K}\Omega \times 2 \, \lambda/4 \, \lambda = 7.5 \text{ K}\Omega$

Input capacitance:

This is made of two parts, the gate capacitance and the poly capacitance. Each input is connected to two transistor gates so

$C_{gate} = 2 \times 1 \text{ fF/}\mu\text{m}^2 \times 2 \, \lambda \times 0.5 \, \mu\text{m}/\lambda \times 4 \, \lambda \times 0.5 \, \mu\text{m}/\lambda = 4 \text{ fF}$

The polysilicon input line has a dimension of $2 \lambda \times 40 \lambda$. Deducting $2 \times 2 \lambda \times 4 \lambda$ of gate area from it, the remaining polysilicon line has a capacitance of

$$C_{poly} = 0.1 \text{ fF/} \mu m^2 \times 2 \lambda \times 0.5 \text{ } \mu m/\lambda \times 32 \lambda \times 0.5 \text{ } \mu m/\lambda = 1.6 \text{ fF}.$$

$$C_{in} = C_{gate} + C_{poly} = 5.6 \text{ fF}.$$

Output capacitance at node F:

There are three parts, the pMOS diffusion, the nMOS diffusion, and the metal wire that connects them.

$$
\begin{aligned}
C_{p\text{-}diff} \quad &= \quad 0.8 \text{ fF/} \mu m^2 \times (4 \lambda \times 0.5 \text{ } \mu m/\lambda \times 6 \lambda \times 0.5 \text{ } \mu m/\lambda) \\
&\quad + 0.3 \text{ fF/} \mu m \times 2 \times (4 \lambda \times 0.5 \text{ } \mu m/\lambda + 6 \lambda \times 0.5 \text{ } \mu m/\lambda) \\
&= \quad 7.8 \text{ fF}
\end{aligned}
$$

$$
\begin{aligned}
C_{n\text{-}diff} \quad &= \quad 0.8 \text{ fF/} \mu m^2 \times (4 \lambda \times 0.5 \text{ } \mu m/\lambda \times 5 \lambda \times 0.5 \text{ } \mu m/\lambda) \\
&\quad + 0.33 \text{ fF/} \mu m \times 2 \times (4 \lambda \times 0.5 \text{ } \mu m/\lambda + 5 \lambda \times 0.5 \text{ } \mu m/\lambda) \\
&= \quad 6.97 \text{ fF}
\end{aligned}
$$

$$
\begin{aligned}
C_{metal} \quad &= \quad 0.05 \text{ fF/} \mu m^2 \times (2 \times 2 + 6 \times 1.5 + 1.5 \times 2) \text{ } \mu m^2 \\
&= \quad 0.8 \text{ fF}
\end{aligned}
$$

$$C_F \quad = \quad C_{p\text{-}diff} + C_{n\text{-}diff} + C_{metal} = 15.57 \text{ fF}$$

In the estimation of C_{metal}, it is important to realize that the diffusion-metal 1 contacts should not be included. In general, the contact capacitance is determined by the material at the bottom of the contact. In this case, the diffusion-metal 1 contacts were accounted for in the estimation of $C_{p\text{-}diff}$ and $C_{n\text{-}diff}$. The reader should be aware that Fig. 4.14 assumes a worst case scenario (minimum depletion region) for diffusion capacitance values. Our C_F value is thus a rather conservative estimation. The real value of C_F is likely to be lower than our estimation above.

Our discussion of capacitance has ignored the capacitance between parallel wires. In sub-micron designs, and especially deep sub-micron designs, the capacitance between two parallel wires becomes significant. This observation is illustrated in Fig. 4.19. As the feature size is getting smaller and smaller, the thickness (i.e., height) of the wires becomes relatively more and more significant. The correct extraction of parasitic capacitance and resistance in deep sub-micron designs is a major research area.

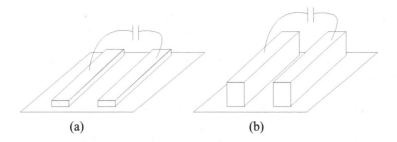

(a)   (b)

Fig. 4.19 Capacitance between two wires: (a) Before sub-micron;
(b) Deep sub-micron.

4.6   RC Timing Model

With the values of resistance and capacitance available for a circuit, the RC timing model provides a very simple approach to estimate its propagation delay. Delays can be found by considering the charging and discharging times of capacitance nodes. Many switch-level simulators use this approach to verify logic function and estimate delay time. Before we demonstrate this approach, we need to introduce a method to estimate the delay of an RC tree. An example RC tree is shown in Fig. 4.20 to explain this method.

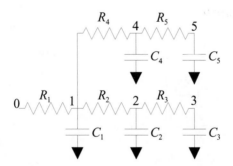

Fig. 4.20 RC tree.

The general formula for estimating the delay from node i to node j in an RC tree is

$$t_{delay} = \sum_{k} R_{ik} C_k \qquad (4.28)$$

where R_{ik} is the total resistance from node i to node k, C_k is the capacitance at node k, and the summation index k goes through all the nodes in the path from node i to node j.

Applying this formula to the RC tree in Fig. 4.20 yields the following expressions:

$$t_{delay}(\text{node } 0 \rightarrow \text{node } 3) = R_1C_1 + (R_1 + R_2)C_2 + (R_1 + R_2 + R_3)C_3$$

$$t_{delay}(\text{node } 0 \rightarrow \text{node } 5) = R_1C_1 + (R_1 + R_4)C_4 + (R_1 + R_4 + R_5)C_5$$

A wire can be considered as a lump element so its parasitic resistance and capacitance values can be found easily according to its dimensions. Alternatively, we can convert it into an RC tree for evaluating its effect on propagation delay. Consider a wire of length l μm and width w μm. To convert this wire into an RC tree, we divide this wire into l segments, each of which has a length of, say, 1 μm. Every segment has a resistance of R_{pu} (resistance per unit) and a capacitance of C_{pu} (capacitance per unit). Now we calculate the RC delay of this wire using two approaches:

Lump Element Approach

$$t_{delay} = R_{pu} \times l \times C_{pu} \times l = l^2 R_{pu} C_{pu} \tag{4.29}$$

RC Tree Approach

$$\begin{aligned} t_{delay} &= R_{pu} \times C_{pu} + 2R_{pu} \times C_{pu} + \cdots + lR_{pu} \times C_{pu} \\ &= \left(\sum_{i=1}^{l} i \right) R_{pu} \times C_{pu} \\ &= \frac{l(l+1)}{2} R_{pu} \times C_{pu} \end{aligned} \tag{4.30}$$

If l is large, then the above expression can be approximated into

$$t_{delay} = \frac{l^2}{2} R_{pu} \times C_{pu} \tag{4.31}$$

which is only half of the delay obtained with the lump element approach. Equation (4.30) is the preferable way to estimate the RC delay of long lines.

Example 4.3

Determine the worst case high-to-low and low-to-high propagation delays at nodes F and Z for the circuit shown in Fig. 4.21.

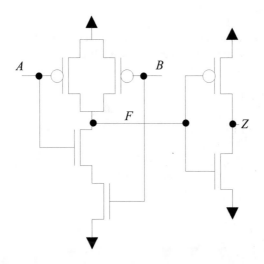

Fig. 4.21 Two-input NAND gate driving an inverter.

High-to-low propagation delay at Z:

Assume that the pMOS and nMOS transistors have effective channel resistance R_p and R_n, respectively. The RC timing model for estimating the high-to-low propagation delay at Z is shown in Fig. 4.22. C_F and C_Z are the total capacitance at nodes F and Z, respectively. Notice that in order to estimate the worst case delay, we assume that only one pMOS transistor in the first stage is turned on.

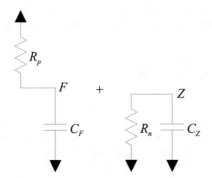

Fig. 4.22   RC model for determining $t_{PHL}(Z)$ and $t_{PLH}(F)$.

From the RC timing model, $t_{PLH}(F) = R_pC_F$, and $t_{PHL}(Z) = R_pC_F + R_nC_Z$, in which the first term R_pC_F is the low-to-high delay from the input to node F and the second term R_nC_Z is the delay between nodes F and Z.

Low-to-high propagation delay at Z:

The RC equivalent circuit for this calculation is shown in Fig. 4.23. Capacitance C_d is the parasitic capacitance between the two pull-down transistors in the first stage of the circuit.

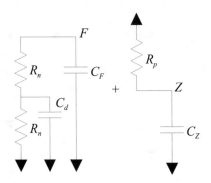

Fig. 4.23   RC model for determining $t_{PHL}(F)$ and $t_{PLH}(Z)$.

From the RC timing model,

$$t_{PHL}(F) = (R_nC_d + 2R_nC_F)$$

$$t_{PLH}(Z) = (R_nC_d + 2R_nC_F) + R_pC_Z.$$

The first two terms in $t_{PLH}(Z)$ constitute the delay from the input to node F ($t_{PHL}(F)$) and the third term is the delay between nodes F and Z.

4.7   Transistor Sizing

The logic function implemented by a CMOS logic circuit depends solely on the way its transistors are connected rather than on the sizes of its transistors. In fact, a circuit consisting of only minimum size transistors will function correctly. However, the performance of such a minimum size design may not meet the design requirements such as speed and logic threshold voltage.

Transistor sizing, which determines the transistor dimensions of a logic circuit, is thus an important step in the design process. The objective of sizing is usually to provide the logic circuit with a desired performance. For example, we may want to provide an equal current driving capability to both the pull-up and pull-down networks so as to equalize t_{PHL} and t_{PLH}. Another possible objective is to select an adequate logic threshold voltage (V_{th}) for a certain application.

In Section 4.2 we have derived the voltage transfer characteristic of the CMOS inverter. Due to different electron and hole mobility values (μ_n and μ_p), the inverter logic threshold V_{th} is located at the center of the logic swing (0 to V_{DD}) only if the pull-up and pull-down transistors are *matched* by widening the pMOS transistor to compensate for its smaller mobility. In other words, the condition of

$$\mu_p \frac{W_p}{L_p} = \mu_n \frac{W_n}{L_n} \qquad (4.32)$$

has to be satisfied for $V_{th} = 0.5 \ V_{DD}$. If $\mu_p = 0.4 \ \mu_n$, also assume that $L_n = L_p$; then W_p has to be sized at 2.5 W_n.[5] Notice that the target of logic threshold voltage adjustment is not always 0.5 V_{DD}. In some applications, it is desirable to move V_{th} closer to one end of the logic swing than to the other (see Problem 4.8).

Another way to look at this sizing issue is that, if both the inverter transistors have the same dimensions, the effective n-channel resistance R_n is smaller than the effective p-channel resistance R_p. If $\mu_p = 0.4 \ \mu_n$, then $R_n = 0.4 \ R_p$. This implies that the low-to-high delay (t_{PLH}) is 2.5 times longer than the high-to-low delay (t_{PHL}). Again, the low-to-high and high-to-low delay times of an inverter can be equalized by widening the pMOS transistor to match with the nMOS transistor. For example, the high-to-low and low-to-high delay times will be identical if W_p is set at $2.5W_n$, assuming $L_p = L_n$.

The sizing of the inverter can be extended to general logic circuits. When a logic circuit with two or more inputs is sized, it is usually for its current driving capability rather than its logic threshold because the latter varies according to the conducting path between V_{DD} and V_{SS} selected by an input combination. However, as a general rule of thumb, the logic threshold of a logic circuit under any input combination should be reasonably close to $V_{DD}/2$ so that a decent noise margin is available at either end of the logic swing.

When a logic circuit has more than one possible pull-up (or pull-down) current path, the idea of "worst case" should be emphasized. In the case of delay time adjustment through transistor sizing, we have to ensure that the worst case current path in the circuit satisfies the requirement. This ensures that the result circuit will work at least as well as it is designed.

The consideration of an equivalent inverter for a logic circuit is helpful in guiding a sizing process. The derivation of an equivalent inverter for a logic circuit is based on the fact that the effective channel resistance of a MOSFET is proportional

---

[5] There is rarely a good reason to lengthen the channel of a transistor unless the objective is to create a transistor to be used as a resistor. The minimum width (2 λ) of a polysilicon wire is the typical channel length.

to L/W, its length to width ratio. Thus, if a number of MOSFETs having length to width ratios of L_1/W_1, L_2/W_3, L_3/W_3, ... are serially connected in a current path, the overall current path resistance will be

$$R = R_s \frac{L_1}{W_1} + R_s \frac{L_2}{W_2} + R_s \frac{L_3}{W_3} + \cdots = R_s \left(\frac{L_1}{W_1} + \frac{L_2}{W_2} + \frac{L_3}{W_3} + \cdots \right) \qquad (4.33)$$

Now, if these serially connected MOSFETs are replaced with a single MOSFET that has a length to width ratio

$$\frac{L_{eq}}{W_{eq}} = \frac{L_1}{W_1} + \frac{L_2}{W_2} + \frac{L_3}{W_3} + \cdots, \qquad (4.34)$$

the current driving capability will be maintained in the equivalent inverter.[6]

Similarly, we can show that the parallel connection of MOSFETs having length to width ratios of L_1/W_1, L_2/W_3, L_3/W_3, ... can be represented by a single MOSFET of

$$\frac{L_{eq}}{W_{eq}} = \frac{1}{\dfrac{W_1}{L_1} + \dfrac{W_2}{L_2} + \dfrac{W_3}{L_3} + \cdots} \qquad (4.35)$$

As an example, two identical MOSFETs with $L/W = 2 \ \lambda/4 \ \lambda$ result in an equivalent L_{eq}/W_{eq} of $4 \ \lambda/4 \ \lambda$ when connected in series and one of $2 \ \lambda/8 \ \lambda$ when connected in parallel.

Example 4.4

Size the transistors in a four-input NAND gate so that its worst case current driving capability is equivalent to that of an inverter with $L_p/W_p = 2 \ \lambda/10 \ \lambda = 0.2$ and $L_n/W_n = 2 \ \lambda/4 \ \lambda = 0.5$. Assume that the channel length of all transistors is fixed at $2 \ \lambda$.

Fig. 4.24 shows the sizing result. In the pull-down network, the equivalent channel length $L_{neq} = 2 \ \lambda + 2 \ \lambda + 2 \ \lambda + 2 \ \lambda = 8 \ \lambda$ and the equivalent channel width $W_{neq} = 16 \ \lambda$ so $L_{neq}/W_{neq} = 0.5$. In the pull-up network, the worst case occurs when only one of four pMOS transistors turns on. In that case, $L_{peq} = 2 \ \lambda$ and $W_{peq} = 10 \ \lambda$ so $L_{peq}/W_{peq} = 0.2$. If more than one pMOS transistor is turned on, the equivalent L_{peq}/W_{peq} would be further reduced to below 0.2.

---

[6] Strictly speaking, this is only true if we ignore the parasitic capacitance between two transistors connected in series.

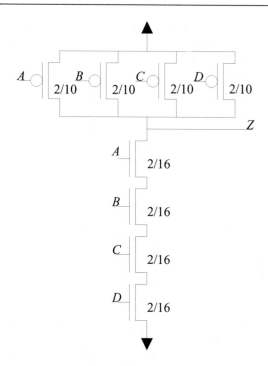

Fig. 4.24 Sizing result of a 4-input NAND gate.

Notice that since the equivalent inverter has matching transistors, its logic threshold voltage is set at $V_{DD}/2$. However, in the NAND gate, this is true only when a single pMOS transistor is considered. The logic threshold voltage will be shifted when more than one pMOS transistor is turned on simultaneously.

Example 4.5

Size the transistors in the circuit of Fig. 4.25 that implements a function $Z = \overline{A + B(C + D)}$ so that its worst case current driving capability is equivalent to that of an inverter with $L_p/W_p = 2\ \lambda/10\ \lambda = 0.2$ and $L_n/W_n = 2\ \lambda/4\ \lambda = 0.5$. Assume that the channel length of all transistors is fixed at $2\ \lambda$.

The transistor schematic for the function Z is shown in Fig. 4.25 with the transistor sizes marked.

The worst current path in the pull-down network is 2 nMOS transistors connected in series (B and C or B and D). Therefore, we select $L_n/W_n = 2\ \lambda/8\ \lambda$ for each of transistors B, C, and D. For transistor A, we select $L_n/W_n = 2\ \lambda/4\ \lambda$ to match with the equivalent inverter. Each of its three pull-down current paths has $L_{eq}/W_{eq} = 0.5$.

Consider the pull-up network. The worst case circuit path has three transistors connected in series (A, C, and D) and thus their sizes are determined to be $L_p/W_p = 2$ $\lambda/30\ \lambda$. The size of transistor B is selected to be $L_p/W_p = 2\ \lambda/15\ \lambda$ so that the current path (A and B) it is in has an equivalent $L_{eq}/W_{eq} = 0.2$. Each of the two pull-up current paths has $L_{eq}/W_{eq} = 0.2$.

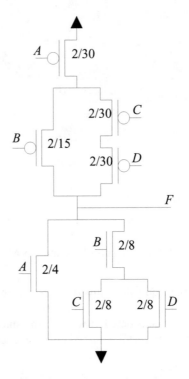

Fig. 4.25 Sizing result of Example 4.5.

4.8 τ-Model

In order to estimate the timing of a circuit, we need to determine all the resistance and capacitance values in the charging/discharging current paths. The transistor gate capacitance and effective channel resistance values are relatively easy to determine since all we need to know is the length and width of the channel. The other parasitic elements such as the diffusion and wiring capacitance values are only available after a layout is created.

In many occasions all we need is a rough timing analysis of the circuit before getting into its layout. A simplified timing model, called the τ-model, can be applied to facilitate the estimation of delay times when a layout is unavailable. The τ-model is based on the enhanced switch transistor model and its associated RC timing model described earlier in this chapter. The only difference is that in the τ-model, only the effective channel resistance and gate capacitance are considered while the diffusion capacitance and wiring effects are ignored.

This simplification apparently has introduced errors into the timing analysis. However, the τ-model is often used to analyze a circuit involving a large capacitance load. In such a case, the load dominates the capacitance so that the diffusion and wiring capacitance can be safely ignored. On the other hand, even in circumstances when this condition cannot be satisfied, the τ-model provides a tool for the designer to gain some insight of the timing relationship in the circuit.

Consider the circuit of Fig. 4.26, in which an inverter built with minimum size transistors drives an identical size inverter. The L/W parameters of the transistors are marked on the schematic diagram.

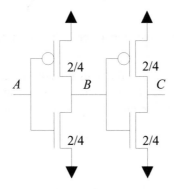

Fig. 4.26 Minimum size inverter driving another identical size inverter.

Let the n-channel resistance (2 λ × 4 λ) be R_n and the p-channel resistance (2 λ × 4 λ) be $R_p = 2.5\ R_n$. Also, let the gate capacitance of a 2 λ × 4 λ transistor be C_g. If we ignore the diffusion capacitance and wiring effect according to the τ-model, we have

$$t_{pHL}(A \rightarrow B) = R_n \times 2C_g$$

$$t_{pLH}(A \rightarrow B) = R_p \times 2C_g = 2.5R_n \times 2C_g$$

We define the average delay time of the driver inverter as

$$t_{avg}(A \to B) = \frac{t_{pLH}(A \to B) + t_{pHL}(A \to B)}{2} = \frac{2R_n C_g + 5R_n C_g}{2} = 3.5 R_n C_g \ (4.36)$$

Now, consider the circuit in Fig. 4.27, which is similar to the one in Fig. 4.26 except that all the transistors are enlarged by doubling their channel widths.[7]

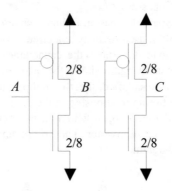

Fig. 4.27 Another example of an inverter driving an identical size inverter.

Doubling the transistor size reduces its effective channel resistance to 1/2 of its original value (R_n or R_p). On the other hand, its gate capacitance has been doubled from its original value (C_g). Therefore, we have for this modified circuit,

$$t_{pHL}(A \to B) = 0.5R_n \times 2 \times 2C_g = 2R_n C_g$$

$$t_{pLH}(A \to B) = 0.5R_p \times 2 \times 2C_g = 5R_n C_g$$

$$t_{avg}(A \to B) = \frac{t_{pLH}(A \to B) + t_{pHL}(A \to B)}{2} = \frac{2R_n C_g + 5R_n C_g}{2} = 3.5 R_n C_g \ (4.37)$$

The significance of this result is that while we have doubled the transistors in the inverter, its average delay time in driving an identical inverter remains a constant. One more example in Fig. 4.28 would really drive this point home. Again we consider an inverter circuit driving an identical size inverter. This time the pMOS transistor is matched with the nMOS transistor.

---

[7] The enlargement of a transistor, unless otherwise stated, refers to the widening of the channel width while keeping the channel length unchanged.

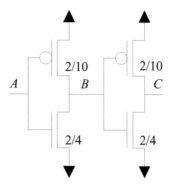

Fig. 4.28 Another example of an inverter driving an identical size inverter.

Due to the matching, both the pull-up and pull-down resistance are R_n. On the other hand, while the nMOS transistor has a gate capacitance of C_g, the pMOS gate capacitance has been increased to $2.5C_g$. The average delay time of the first inverter with the second one as a load is determined:

$$t_{pHL}(A \rightarrow B) = R_n \times (C_g + 2.5C_g) = 3.5 R_n C_g$$

$$t_{pLH}(A \rightarrow B) = R_n \times (C_g + 2.5C_g) = 3.5 R_n C_g$$

$$t_{avg}(A \rightarrow B) = \frac{t_{pLH}(A \rightarrow B) + t_{pHL}(A \rightarrow B)}{2} = 3.5 R_n C_g \qquad (4.38)$$

Once again, the average delay time of the first inverter driving an identical inverter remains unchanged. This constant delay time $3.5R_nC_g$, which is defined as 1 τ, can be considered as a pseudo physical parameter of a fabrication process. We summarize the result of the above discussion into the following statement: If we keep the channel lengths at their minimum dimension (i.e., 2 λ), ignore the diffusion capacitance and wiring effect, then an inverter driving an identical inverter has an average delay time of $\tau = 3.5R_nC_g$ which is independent of the actual transistor sizes.

The application of the τ-model will be demonstrated in the following examples. Let us define a unit (1×) inverter as the minimum size inverter (i.e., with 2 λ × 4 λ transistors). An n× inverter is then a unit inverter enlarged n times (i.e., with 2 λ × 4n λ transistors).

Example 4.6

Fig. 4.29 shows a 1× inverter driving a 10× inverter. Find the average delay time of the driving inverter.

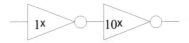

Fig. 4.29 1× inverter driving a 10× inverter.

According to the τ-model, the average delay time of the driving 1× inverter is 10 τ. This is due to the fact that a 10× inverter has ten times the input capacitance of a 1× inverter.

Example 4.7

Find the average delay time of the driving inverter in Fig. 4.30.

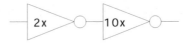

Fig. 4.30 2× inverter driving a 10× inverter.

Since we increase the driving inverter into 2×, the channel resistance of its transistors will be reduced to $0.5R_n$ and $0.5R_p$. The average delay time of the 2× inverter is thus also reduced to 5 τ.

Example 4.8

In Fig. 4.31, the 1× driving inverter has a fan-out of two 10× inverters. Find the average delay time of the driving inverter.

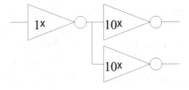

Fig. 4.31 1× inverter driving two 10× inverters.

The τ-model can also be applied to estimate the delay time when the fan-out is larger than one. From the viewpoint of load, this is equivalent to having a 20× inverter connected and the average delay time of the 1× driving inverter is thus 20τ.

It should be noted that the τ-model, which has been described using inverters as examples, can be readily extended to other complementary logic circuits.

4.9   Driving Large Loads

The usefulness of the τ-model is evident in cases when a CMOS circuit is used to drive a large capacitance load which has a value that is many times larger than the input capacitance of an inverter. Consider the non-inverting buffer in Fig. 4.32. Both inverters are of the same size (1×). Assume that $C_L \gg 2C_g$, where C_g is the 2 $\lambda \times 4 \lambda$ gate capacitance.

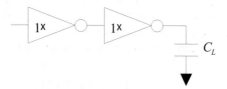

Fig. 4.32 Non-inverting buffer driving a large load.

The average delay time of the non-inverting buffer can be estimated easily using the τ-model. Assume that the 1× inverter has an input capacitance of C_{in}. If we substitute the load C_L with an inverter that has an input capacitance of C_L, the average delay of the non-inverting buffer will not be affected at all. This is shown in Fig. 4.33. The inverter replacing the load will be a (C_L/C_{in})× inverter and thus the average delay of the non-inverting buffer is $(1 + (C_L/C_{in}))$ τ.

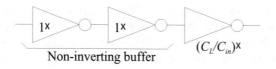

Fig. 4.33 Replacing a large load with a (C_L/C_{in})× inverter.

To appreciate the sizing of the buffer for delay time adjustment, let us assume that $(C_L/C_{in}) = 100$ and thus the average delay of the buffer driving the load C_L is 101 τ. If this is considered to be too slow for a certain application, the second inverter in the buffer can be enlarged to provide more current to charge/discharge C_L more rapidly. For example, we replace the second inverter with a 100× inverter as shown in Fig.

4.34; then the average delay time of the second inverter will be reduced from 100 τ to only 1 τ. However, this simple solution does have a practical problem. The 100× inverter, which has a shorter delay time itself, presents an input capacitance that is 100 times of a 1× inverter. Therefore, the first inverter will have a 100 times longer delay time. The average delay of the entire buffer thus remains unchanged.

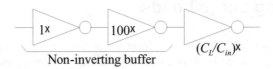

Non-inverting buffer

Fig. 4.34 Replacing the second inverter with a 100× inverter.

An optimal size, say, $k\times$, of the second inverter to minimize the buffer delay can be determined as follows. We assume that the first inverter is unchanged so that the effect of modifying the buffer will not propagate to the stage driving the buffer. The average delay of the buffer can be expressed as a function of k:

$$t_{avg}(\text{buffer}) = \left(k + \left(\frac{C_L}{C_{in}} \right) \left(\frac{1}{k} \right) \right) \tau , \qquad (4.39)$$

which is minimized when $k = \sqrt{\dfrac{C_L}{C_{in}}}$.

In our numerical example when $(C_L/C_{in}) = 100$, k should be set at 10. The average delay of this buffer is merely 20 τ.

In order not to overdesign the buffer, when the goal is to keep the delay at or below a specified value other than the minimum delay time, equation (4.39) can also be used to determine k. On the other hand, if the minimum delay time achievable by the buffer is still too slow, more stages can be used. In Fig. 4.35, we use n inverters, each of which is a times larger than its previous one, to drive C_L. Again, in order not to affect the stage before the buffer, the first inverter is kept as a 1× inverter.

C_{in} is the input capacitance of the 1× inverter and $C_L = a^n C_{in}$. In order to minimize the overall delay time, there are two parameters to be determined in this buffer, the number of stages n and the stage enlargement a. The average delay time of the buffer is

$$t_{avg} = na\tau \qquad (4.40)$$

From the relationship $C_L = a^n C_{in}$, we have

$$n = \frac{\ln\left(\dfrac{C_L}{C_{in}}\right)}{\ln a} \, . \tag{4.41}$$

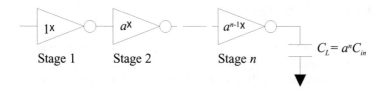

Fig. 4.35 Chain of inverter stages driving a large load.

Substitute n into equation (4.40),

$$t_{avg} = \ln\left(\frac{C_L}{C_{in}}\right)\frac{a}{\ln a}\,\tau \tag{4.42}$$

which has a minimum when $a = e$, the natural number ($e \approx 2.718$). If $a = e$,

$$n = \ln\left(\frac{C_L}{C_{in}}\right) \tag{4.43}$$

and the optimal delay time achievable by the technology is obtained. In practice, a is chosen to be an integer between 2 and 10.

Again, the designer should not overdesign for unnecessary speed and the above method can be alternatively used to determine the *smallest* buffer that can satisfy a given delay constraint.

Example 4.9

The equivalent input capacitance C_{in} of an inverter is .01 pF and the delay time when its load is an identical size inverter is 1.0 ns. This inverter is used to drive an output pin with a capacitance of $C_L = 11$ pF. For a minimum average delay time, how many buffering inverters should be used? What is the minimum average delay time?

Assume that the size of the first inverter in the buffer is kept unchanged. To achieve the minimum average delay time, $n = \ln(11/.01) \approx 7$ stages. So 6 additional inverters should be added between the original inverter and C_L. Notice that in this case since an even number of inverters were added to the circuit, no change to the logic is needed.

Selecting $a = 3$, the average delay of the circuit including the original inverter is estimated to be $7 \times 3 \times 1.0$ ns $= 21$ ns.

4.10 Power Dissipation

Power dissipation is an important evaluation criterion of integrated circuits. For battery-operated systems, a high power dissipation reduces the battery life. The problem of heat dissipation will also have to be dealt with. CMOS complementary logic circuits are known for their low power dissipation. We use an inverter as an example to pursue the following power analysis for a complementary logic circuit.

When the CMOS inverter is in a steady state, regardless of whether the output is a 1 or a 0, only half of the circuit is conducting and there is no closed path between V_{DD} and V_{SS}. Since no current flows, the steady state power is 0 barring the small leakage current.

When the CMOS inverter is in transition, we have seen in Fig. 4.5 that both transistors are turned on so a current flows between V_{DD} and V_{SS} through them. This is called the short-circuit current, which contributes to the dynamic power dissipation in the CMOS inverter. A well-designed circuit operating with well-behaved signals of reasonably fast rise time and fall time would go through the transition quickly. The short-circuit power dissipation is thus less significant comparing to the dynamic power caused by the current that flows through the transistors to charge or discharge a load capacitance C_L.

In the following derivation of an expression for estimating the dynamic power dissipation, we assume that the input signal stays unchanged long enough for the output voltage across C_L to reach its final value. We will see that this assumption is satisfied by the basic principle of designing a clock signal. The clock period should be long enough to allow all transitions to complete.

When C_L is being charged through a pMOS transistor with effective channel resistance R_p, as depicted in Fig. 4.36, the voltage $v(t)$ across C_L rises from 0 to V_{DD}.

The current flowing through the capacitance C_L and the voltage across it are

$$i(t) = \frac{V_{DD}}{R_p} e^{-t/R_p C_L} \tag{4.44}$$

and

$$v(t) = V_{DD}(1 - e^{-t/R_p C_L}) \tag{4.45}$$

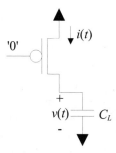

Fig. 4.36 Charging a load capacitance through a pMOS transistor.

The charge current $i(t)$ comes from the power source V_{DD} and becomes the charge accumulated on C_L. The energy drawn from the power source during the charging period is thus

$$\int_0^\infty V_{DD}i(t)dt = V_{DD}\int_0^\infty i(t)dt = V_{DD}Q = V_{DD}C_LV_{DD} = C_LV_{DD}^2 \qquad (4.46)$$

where $Q = C_LV_{DD}$ is the charge accumulated on the capacitance C_L. The lower and upper bounds of the integration in (4.22) are the result of assuming that the current $i(t)$ will flow until it stops when $v(t)$ reaches its final value V_{DD}.

At the end of the charging interval, the capacitor has the energy of

$$\int_0^\infty i(t)v(t)dt = C_LV_{DD}^2(e^{-t/R_pC_L} - 0.5e^{-2t/R_pC_L})\Big|_0^\infty = 0.5C_LV_{DD}^2 \qquad (4.47)$$

stored on it. Therefore the energy of $0.5C_LV_{DD}^2$ is dissipated in the pMOS transistor.

On the other side of the transfer curve, when C_L is discharged through the nMOS transistor, the voltage across it drops from V_{DD} to 0. The energy of $0.5C_LV_{DD}^2$ is thus dissipated in the nMOS transistor.

Summarizing the operation of a CMOS inverter, $0.5C_LV_{DD}^2$ of energy is dissipated when the output switches from 0 to 1 and another $0.5C_LV_{DD}^2$ of energy is dissipated when the output switches from 1 to 0. Assume that the inverter is driven by a free running clock of frequency f. This represents the worst case operating scenario in which the inverter switches twice in a clock cycle. The average dynamic power dissipation due to charging and discharging capacitance C_L is

$$P_D = fC_LV_{DD}^2 \qquad (4.48)$$

It can be seen from (4.48) that the operating frequency of the circuit determines its power dissipation. This relationship allows us to perform a tradeoff between speed and power dissipation. Increasing the speed and thus the operating frequency increases power dissipation. The term $C_L V_{DD}^2$ in (4.48), which is called the delay-power product, can be used as a figure of merit to evaluate VLSI technologies.

In a CMOS circuit all switching occur more or less simultaneously in a synchronous (clocked) circuit. This results in a large surge in the power supply current when the clock ticks. If the power supply cannot keep up with the demand of current, glitches and spikes occur in the signals, which may affect the normal operations of the circuit. The inductance of the package (pins, bond wires, etc.) may present an obstacle to the supply of sufficient current. This problem can be treated by incorporating bypass capacitors on chips. Typically the bypass capacitance is around ten times of the total capacitance involved in the switching.

4.11 Latch-Up

The latch-up problem is probably the most significant drawback of CMOS circuits. Fig. 4.37 shows a cross section of a CMOS structure illustrated with two unexpected guests: two bipolar transistors. The pnp bipolar transistor consists of the source/drain of the pMOS transistor, the n-well, and the substrate. The npn bipolar transistor is formed of the source/drain of the nMOS transistor, the substrate, and the n-well. The resistance of the n-well and the substrate are represented by R_w and R_b, respectively.

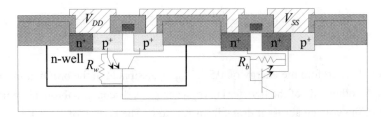

Fig. 4.37 Parasitic bipolar transistors in a CMOS structure.

The bipolar transistors and the well and substrate resistance are extracted and modeled in Fig. 4.38. Normally the p-n junctions of the transistors are reverse-biased so the parasitic bipolar transistors are not conducting.

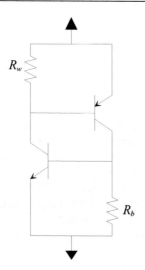

Fig. 4.38 Parasitic transistors of a CMOS logic.

Latch-up is caused by an accidentally forward-biased p-n junction. This could be the junction between the pMOS diffusion and the n-well, or the one between the nMOS diffusion and the substrate. The p-n junctions in a CMOS circuit could be forward-biased for a number of reasons. For instance, the drain, being the output of the circuit, may receive a so-called "inductive kick" and goes beyond V_{DD}. Forward-biased junctions may also arise when an external energy source, such as radiation or light, causes unwanted free electrons and holes to flow in the structure, which may turn on the parasitic bipolar transistors.

Let us suppose the drain of the pMOS transistor is temporarily raised beyond V_{DD}. The p-n junction between the drain and the n-well is forward biased and the parasitic pnp transistor conducts and pulls current from V_{DD} to V_{SS}.

The current created by the conducting pnp transistor causes a base-emitter voltage drop on the substrate resistance R_b. If this voltage drop is large enough to forward bias the p-n junction between the substrate and the nMOS transistor drain, the npn transistor conducts and causes more current to flow through the structure. This current creates a base-emitter voltage drop over the well resistance R_w. A large enough voltage drop will maintain the conducting of the pnp transistor even though the original cause of forward-bias current may now have disappeared.

The pnp transistor draws current to support the conducting of the npn transistor and vice versa. A latch-up has occurred. Latch-up may disrupt the CMOS circuit normal operation and can cause large currents to flow and destroy the devices.

A number of methods can be used to prevent latch-up from occurring. The values of R_w and R_b should be reduced as much as possible, thereby lowering the voltage drops across them. The gains of the parasitic bipolar transistors should be reduced by increasing the spacing between the nMOS and pMOS transistors. This results in the most conservative spacing rule in the design rules. N-diffusion and p-diffusion areas are separated by at least 10 λ.

4.12 Summary

In this chapter we have provided a review of MOSFET characteristics. Our purpose was to provide an appropriate background for the simulation of CMOS circuits. After giving a brief introduction to SPICE simulation, a MOSFET resistive switch model was created for the estimation of propagation delay. The calculation of parasitic resistance and capacitance values in CMOS circuits was described. The use of a τ-model to gain some insight of CMOS propagation delay and to design buffers for large capacitance loads was discussed. A way to estimate CMOS power dissipation was described. The CMOS latch-up problem was explained and its prevention provided.

4.13 To Probe Further

MOSFET and CMOS Electronics:

*   S. Sedra and K. C. Smith, *Microelectronic Circuits*, 4th *Edition*, Oxford University Press, 1998.

*   K. Lee, M. Shur, T. A. Fjeldly, and T. Ytterdal, *Semiconductor Device Modeling for VLSI*, Prentice-Hall, 1993.

SPICE:

*   G. W. Roberts and A. S. Sedra, *SPICE*, Oxford University Press, 1997.

*   S. Sedra and K. C. Smith, *Microelectronic Circuits*, 4th *Edition*, Oxford University Press, 1998.

SPICE Models:

*   http://www-device.eecs.berkeley.edu/~bsim3/

RC Circuits:

*   D. E. Scott, *An Introduction to Circuit Analysis: A Systems Approach*, McGraw-Hill, 1987.

*   L. P. Huelsman, *Basic Circuit Theory*, 3rd *Edition*, Prentice-Hall, 1991.

4.14 Problems

Assume that the 1 μm technology with the values given in Fig. 4.14 is used in the following problems.

4.1   A polysilicon wire with a length of 80 λ and a width of 4 λ is used to connect the output of a two-input NAND gate to an inverter input and a NOR gate input. All pull-up transistors have W_p = 6 λ, L_p = 2 λ and all pull-down transistors have W_n = 3 λ, L_n = 2 λ. Solve the following problems by considering only gate capacitance, effective channel resistance, and the resistance and capacitance of the polysilicon wire.

(a)   Draw equivalent circuit(s) for evaluating the worst case high-to-low and low-to-high delay times at the input of the inverter. Label the R's and C's clearly and compute their values.

(b)   Change the nMOS transistor dimensions of the NAND gate so that its high-to-low delay time is less than 0.05ns.

(c)   Change the pMOS transistor dimensions of the NAND gate so that the low-to-high delay time is less than 0.05ns.

(d)   Estimate the total power dissipation of this circuit if the inputs of the NAND gate has a maximum switching rate of 50 MHz. Use the transistor sizes found in parts (b) and (c).

4.2   This problem refers to the layout diagram given in problem 3.5 in Chapter 3. Assume nodes B and D are now connected by a metal wire and determine worst case t_{PHL} and t_{PLH} at node E.

4.3   Consider the following structures:
a)   A diffusion line with a width of 4 λ.
b)   A polysilicon line with a width of 2 λ.
c)   A metal 1 line with a width of 4 λ.
d)   A metal 2 line with a width of 4 λ.
Determine in each case the longest length of the structure that can be ignored in propagation delay evaluation. (Hint: Apply the engineer's rule of thumb: "A term that contributes less than 10% to the final value can be ignored in the calculation.")

4.4   This refers to the problem described in Example 4.9. Now the technology advances and this chip can be scaled down in size by a factor of 8. Assume that the equivalent inverter input capacitance and delay time are scaled down with the size linearly. Should the designer increase or reduce the number of inverters in the chain? By how many? Will the minimal delay time to drive this load be shorter or longer? By how much?

4.5   Design an output buffer to be used to drive an external load of 50 pF. The output stage that uses this buffer has $W_p = W_n = 4$ µm, $L_n = L_p = 1$ µm. Try to find the fastest buffering scheme.

4.6   Redo problem 4.5 so that it will produce a delay of less than or equal to 0.4 ns. Use resources wisely and do not overdesign your buffer. In other words, do not design an overly speedy buffer.

4.7   Calculate the average current drawn by 32 of the buffer created in problem 4.6 if they are used in a system with a clock rate of 200 MHz. If the peak current is approximately 2 times the average current, what is the size of the power rails (V_{DD} and V_{SS}) required for these buffers if the allowable current density of the metal layer is 1 mA/µm?

4.8   Determine the logic threshold voltage of an inverter if (a) $R_n = 0.4\ R_p$; (b) $R_n = R_p$; and (c) $R_n = 2\ R_p$.

4.9   A polysilicon line has a maximum power density of 20 Watts/cm². It is used in a circuit with $I_{max} = 1$ mA. What would be the smallest allowed dimensions of a polysilicon resistor with 1000 Ω?

4.10   Use the following parameters to carefully plot the voltage transfer function of an inverter:
       $V_{DD} = 5$ V , $W_p = L_p = W_n = L_n = 2.5$ µm, $V_{tp} = -1$ V, $V_{tn} = 1$ V, $t_{ox} = 0.05$ µm. For any other parameters, use typical values given in this chapter.

4.11   Repeat 4.10 but with $W_p = 10$ µm. The other parameters remain unchanged.

4.12   Describe the effects of scaling on circuit performance: area, speed, and power. Assume that all dimensions of an IC, including those vertical to the substrate, are scaled by dividing them by a constant factor α. Also assume that the voltage (V_{DD}) is scaled down by dividing by the same factor α.

4.13   Repeat Problem 4.12 but assume that the voltage (V_{DD}) remains unchanged.

4.14   Determine τ for an inverter with $L_n = L_p = 2$ µm, $W_n = W_p = 4$ µm.

4.15   Determine τ for an inverter with $L_n = L_p = 4$ µm, $W_n = W_p = 8$ µm.

4.16   Compare the results of Problems 4.14 and 4.15 and comment on the following statement: "If we consider only effective channel resistance and gate capacitance, for a given technology, the average delay time of an inverter driving an identical inverter is a constant τ."

4.17 All the transistors used in a multiple level circuit have dimensions of $W = 6$ μm and $L = 6$ μm. This circuit has a delay of 50 ns when its load is a minimum size inverter (i.e., $W = 4$ μm and $L = 2$ μm). Consider only channel resistance and gate capacitance in the following problems. If the load is replaced by a 50 pF capacitor, what is the delay time?

4.18 If two inverters are used to form a non-inverting buffer between the circuit and the 150 pF capacitor, determine the dimensions of the transistors used in these two inverters so that the delay time from the circuit input to the 150 pF capacitor is minimized.

4.19 The τ-model is used to design the buffering problem. The 1× inverter is made of minimum size transistors. Determine f and g so that the total delay of this circuit is minimized.

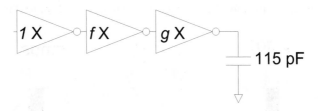

Fig. 4.39   Buffer design problem for Problem 4.19.

4.20 Use a layout editor to create a CMOS 4-input composite gate ($Z = \overline{A(B+C)D}$). Use transistors with $L = 2$ λ, $W = 8$ λ. Perform a switch level simulation of your composite gate. Perform a circuit level simulation of your composite gate using SPICE.

4.21 Use your layout and simulation result from Problem 4.20 to derive the effective channel resistance values and load capacitance values for the technology used in your SPICE simulation.

4.22 All the transistors used in the circuit shown in Fig. 4.40 have $W = 4$ λ, $L = 2$ λ. The part represented by the box (i.e., from the left edge to the right edge of the box labeled as "Circuit") has a delay of $T\tau$ (the delay of the "Buffer" included). τ is the average delay time of an inverter driving an identical inverter. What is the overall average delay time in terms of τ as a function of α when the Driver is an $\alpha×$ inverter.

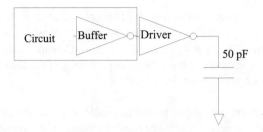

Fig. 4.40 Circuit for Problem 4.22.

4.23   Fig. 4.41 shows a partial layout of an inverter with minimum dimensions (not drawn to the scale). Mark clearly the important dimensions (in λ's) on the layout according to the scmos technology. Estimate the following values: the input capacitance for this inverter; the drain capacitance for the transistors; the output capacitance for this inverter; the channel resistance for this inverter.

Fig. 4.41 Layout for Problem 4.23.

Chapter 5 Sequential Logic Circuits

The ability to remember ...

The design and analysis of CMOS combinational logic circuits have been explained in Chapters 2-4. A combinational logic circuit does not have memory so its operation history has no effect on its behavior. The output of a combinational circuit is thus completely determined by its present input. This simplicity limits the applications of combinational logic circuits.

A television (TV) control is an excellent example to demonstrate the usefulness of the ability to remember. When the user presses the "up" button on the TV remote control, the TV responds and switches to the next available channel. Suppose the TV was tuned to, say, Channel 5 before the button was pressed; it would now be tuned to Channel 6. The user can press the "up" button again to tune the TV to Channel 7 and so forth. In order for the TV to operate correctly, it must be aware of its present channel to determine the next channel destination when the user presses the "up" button. A combinational logic circuit cannot handle this type of application.

We need another type of logic circuits that we call sequential logic circuits. The present state of a sequential logic circuit, which is determined by the operation history of the circuit, is stored in its memory. The output of a sequential logic circuit is produced according to its input and its state. Sequential logic circuits can solve numerous problems that are beyond the capability of combinational logic circuits.

The TV control described above is apparently a sequential circuit. The present state of the circuit is the channel to which the TV is tuned. Whenever the button on the TV remote control is pressed (the input), the circuit decides the new channel (the next state). A state transition then occurs. The TV is then tuned to the desired channel (the output).

In this chapter, we discuss the design and analysis of CMOS sequential logic circuits.

5.1 General Structure

We begin by reviewing the general structure of a sequential logic circuit. A sequential logic circuit can operate with or without the control of a clock signal. A clock signal is a periodic square wave that can be used to trigger state transitions. Fig. 5.1 shows an example of a clock signal which has a cycle time (period) of T_{cyc}. The clock frequency is thus

$$f = \frac{1}{T_{cyc}}. \tag{5.1}$$

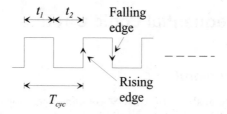

Fig. 5.1 Clock signal.

Each clock cycle is divided into two parts, t_1 and t_2. The clock signal level is high during t_1 and is low during t_2. The duty cycle of a clock signal is defined as

$$\text{Duty Cycle} = \frac{t_1}{t_1 + t_2}. \qquad (5.2)$$

In addition to its frequency and duty cycle, a clock signal is qualified by considering its rise time and fall time. Fig. 5.2 shows the definitions of rise time and fall time. Rise time is the time the clock signal takes to go up from 10% to 90% of its high value. Fall time is the time the clock signal takes to go down from 90% to 10% of its high value. A good clock signal must have reasonably fast rise time and fall time (typically below 1/10 of the clock cycle time).

One can use the levels (high and low) or the transitions (rising edge and fall edge) of a clock signal to synchronize the state transitions in a sequential logic circuit. A sequential logic circuit that utilizes a clock signal is called a synchronous sequential logic circuit. A sequential logic circuit that is not synchronized by a clock signal is said to be asynchronous. A brief introduction to asynchronous sequential logic is provided in the next section. This serves as the background of designing a D-latch, which is the basic CMOS memory element. The rest of the chapter then concentrates on the design and analysis of synchronous sequential logic circuits.

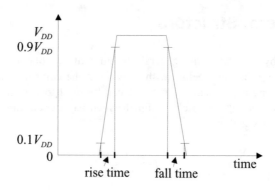

Fig. 5.2 Rise time and fall time.

Fig. 5.3 shows the general structure of a synchronous sequential logic circuit. The core of the circuit is a combinational logic circuit which accepts inputs (*X* and *y*) and produces outputs (*Z* and *Y*).[1] The output (*Y*) is stored in the memory as a state variable. The number of bits in the state variable decides the number of available states; so a sequential logic circuit is also called a finite state machine. A clock signal triggers the capturing of state variables into the memory. The memory typically consists of either level-triggered latches or edge-triggered flip-flops. We will discuss the CMOS implementation of latches and flip-flops shortly.

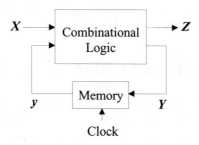

Fig. 5.3   General structure of a sequential logic circuit.

In Fig. 5.3, *X* and *Z* are referred to as the input and output of the sequential logic circuit, respectively. On the other hand, *Y* and *y* represent the next and present states of the circuit, respectively. Two types of synchronous sequential circuits can be created within this general structure — the Mealy machine and the Moore machine. These two sequential circuit structures differ mainly in the ways they generate their outputs.

The Mealy machine uses both its input and present state to generate its output. The combinational circuit of a Mealy machine can be expressed as

$$Z = f(X, y)$$
$$Y = g(X, y), \tag{5.3}$$

where $f(\bullet)$ and $g(\bullet)$ represent logic functions.

In contrast, the Moore machine uses only its present state to determine its output. The combinational circuit of a Moore machine can be expressed as

$$Z = f(y)$$
$$Y = g(X, y). \tag{5.4}$$

---

[1] Bold faced letters indicate a group of associated signals.

A further discussion of these machines is beyond the scope of this book. Readers who need a review of this subject can consult one of the digital logic design books listed at the end of the chapter. We will only point out that a Mealy machine can be converted into an equivalent Moore machine and vice versa. It can be seen from (5.4) that all output changes in a Moore machine are synchronized to the clock, since output Z is a function of the state only; so it can change only when the state changes. In other words, output Z remains stable during input changes and produces no unwanted glitches.[2] On the other hand, a Mealy machine has the advantage of offering the designer more flexibility in designing output (Z) and state variable (Y) functions, since its output is a function of both its input and state. A Mealy machine usually needs fewer states than its equivalent Moore machine.

5.2 Asynchronous Sequential Logic Circuits

Memory elements provide a temporary storage of state variables to be used at a later time. The basic memory element employed in a CMOS sequential logic circuit is the D-latch. While this chapter mainly deals with synchronous sequential circuits, the D-latch is nevertheless an asynchronous sequential circuit.

In order to gain some insight into the design and use of a D-latch, we begin by discussing the general structure of an asynchronous sequential circuit. The general structure of an asynchronous sequential circuit is shown in Fig. 5.4.

Compare Fig. 5.3 and Fig. 5.4. An asynchronous sequential circuit feeds the next state, Y, directly back to y, without going through memory elements. The lack of memory elements and their controlling clock signal implies that the timing of state transitions is determined solely by the combinational logic delay.

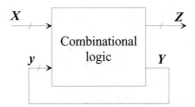

Fig. 5.4 General structure of an asynchronous sequential circuit.

Suppose we change X at time t. The state variable Y responds and changes at time $t + \Delta t$. Since Y is immediately reflected at y, its change could affect the unfinished operation of the circuit and result in an incorrect circuit behavior.

---

[2] Glitches cause the circuit to switch unnecessarily and thus consume power.

The propagation delays of individual bits in the state variable Y are thus critical to the correct operation of the circuit. A number of reasons, such as manufacturing and temperature variations, can cause these delays to vary from their nominal values. A change in the responding order of the state variable bits may, as we will show below, lead the circuit down different transition paths into different states. This is called a state transition hazard.

Consider that a sequential logic circuit is currently at state $y = 000$. The circuit is designed to respond to a certain input X and produce a new state $Y = 011$. The transition between 000 and 011 involves the changes of two state variable bits. Depending on the combinational logic delay, Y may go through one of two intermediate states (010 or 001) before it reaches its final state (011). In the case of a synchronous sequential logic circuit, one can adjust the clock speed to capture the new state only after all state variable bits settle down so that delay variations cause no hazard. An asynchronous sequential circuit, in contrast, may respond differently to intermediate states 010 and 001 since they go right back to the input of the combinational logic circuit. The designer must ensure that any possible intermediate state will eventually evolve into the desired final state to avoid a state transition hazard.

Consider the asynchronous sequential circuit shown in Fig. 5.5, in which the output of an inverter is connected to its input. A simple analysis illustrated in Fig. 5.6 shows that this circuit cannot settle into a stable state. In other words, the circuit oscillates.

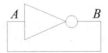

Fig. 5.5 Inverter connected into an asynchronous sequential logic circuit.

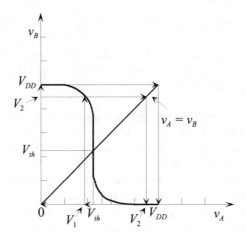

Fig. 5.6 Oscillation of an inverter with its output connected to its input.

Fig. 5.6 shows the ideal voltage transfer characteristic of an inverter. A straight line with unity slope (i.e., $v_A = v_B$) is also included in Fig. 5.6. This straight line represents the relationship that the voltage at A (v_A) equals the voltage at B (v_B). The straight line intersects the voltage transfer curve at $v_A = v_B = V_{th}$ (V_{th} is the logic threshold voltage of the inverter). While this intersection represents a theoretically stable operating point, in practice the circuit cannot operate at this point for any period of time.

The reason is that interference or noise is inevitably present in any circuit. Assume that the circuit starts at $v_A = v_B = V_{th}$. Let v_A decrease by a small amount to V_1. The voltage at the output of the inverter will increase to V_2. This results in $v_A = V_2$ which causes output v_B to go to 0. Now $v_A = 0$ so v_B responds by going to V_{DD}. The change of v_A from 0 to V_{DD} produces a v_B that is 0. From this point on, the circuit output oscillates between 0 and V_{DD} with a frequency determined by the inverter delay. A trace showing the occurring and sustaining of this oscillating process is shown in Fig. 5.6.

In the description above, we assumed an initial voltage decrement at v_A. A similar result would have been produced had we instead assumed a voltage increment at v_A. The reader is encouraged to verify this (see Problem 5.4).

The unstable asynchronous sequential logic circuit shown in Fig. 5.5 is called a ring oscillator. As noted, the inverter delay determines the oscillation frequency. In practice, a ring oscillator is often formed by connecting an odd number of inverters into a loop. The more inverters used in the loop, the lower is its oscillating frequency.

The ring oscillator is often used as a relatively simple means for measuring the inverter propagation delay. The average delay of each inverter in the ring oscillator is calculated as T_{cyc}/n, where T_{cyc} is the oscillation period and n is the number of inverters in the loop. A ring oscillator made of five inverters is shown in Fig. 5.7.

Fig. 5.7 Ring oscillator consisting of 5 inverters.

A ring oscillator can be used as a low-cost, low-quality clock generator. However, the oscillating frequency of a ring oscillator, being a function of the inverter delay, is sensitive to manufacturing and temperature variations.

5.3   D-Latch

A memory element can be built by setting up an asynchronous sequential logic circuit that has self-sustaining stable states. A stable state of an asynchronous sequential circuit occurs when no further changes are possible. Refer to Fig. 5.4; if we keep the input X unchanged, the circuit is in a stable state if and only if

$$Y(t + \Delta t) = Y(t). \tag{5.5}$$

A CMOS memory element that employs this principle is shown in Fig. 5.8, in which two inverters are connected into a non-inverting and self-sustaining loop.

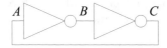

Fig. 5.8 Non-inverting loop formed by two inverters.

The circuit in Fig. 5.8 provides a non-inverting function from A to C so its voltage transfer curve (v_C vs. v_A) takes the shape in Fig. 5.9. The unity slope straight line represents the relationship that the voltages at A (v_A) equal the voltage at C (v_C) since A and C are connected together. This figure can be analyzed in a manner similar to what we did in Fig. 5.6. There are three intersection points between the voltage transfer curve and the straight line, indicating three operating points. The center operating point is unstable. A small increment in v_A will cause the circuit to shift toward $v_A = V_{DD}$ which is a stable operating point. On the other hand, if there is a small decrement in v_A, the operation of the circuit goes toward $v_A = 0$ and stays there. If we consider v_A as the state of the circuit, it can stay at either 1 or 0; so the circuit has two stable states.

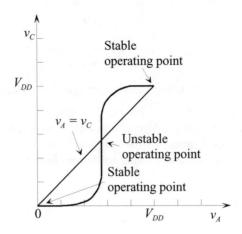

Fig. 5.9 Operating points of a CMOS memory element.

Now we have a bistable CMOS memory element. In order to make it useful, we must provide a way to set its state to a desired value. Fig. 5.10 shows a D-latch that implements this function. A pass-transistor is used to feed an externally applied signal (D) into the loop when its control signal L equals 1. In order to avoid a fighting condition, in which a node is driven by two potentially opposite signals (D and Q), to occur when L equals 1, another pass-transistor is provided to selectively open and close the feedback loop. This structure can also be viewed as that the two pass-transistors form a 2-to-1 multiplexer, which is used to feed either input D or output Q as the input of the first inverter.

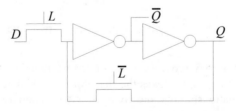

Fig. 5.10 A D-latch.

Fig. 5.11 shows an alternative way of constructing a D-latch, which replaces the nMOS pass-transistors driven by $\overline{L}$ with a pMOS transistor so that no inverted control signal is needed. The pass-transistors in a D-latch can also be replaced by transmission-gates to avoid the occurrence of weak signals.

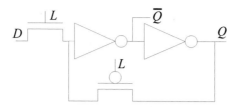

Fig. 5.11 An alternative D-latch utilizing a pMOS transistor.

The behavior of the D-latch is illustrated in the timing diagram of Fig. 5.12. Initially L is set to 0 so the inverter loop is closed and isolated from input D. Output Q has an initial value which is unknown to us; so it is indicated in the timing diagram as a shaded area. When L is changed to 1, input D passes through the inverters and appears at output Q after the inverter delays (not shown in the timing diagram). If input D is changed when $L = 1$, output Q follows. The D-latch is thus said to be transparent during the time when L equals 1. The last value of D before L is changed to 0 is captured by the latch and retained at output Q until L is brought up to 1 again.

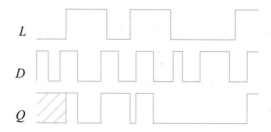

Fig. 5.12 Behavior of the D-latch.

The use of D-latches as memory elements in a sequential circuit is demonstrated in Fig. 5.13. Each bit of the state variable is stored in a D-latch.[3] We have seen that the D-latch responds to the level of control signal L so it is referred to as a level-controlled device. This characteristic causes a few challenges in the use of D-latches in a sequential circuit. We first explain the issues of using D-latches. In a moment we will discuss the development of a more sophisticated memory element — the D-flip-flop, which captures its data in response to the transitions (rising and falling edges) of a clock signal.

---

[3] Only a single representative D-latch is shown in Fig. 5.13.

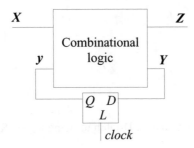

Fig. 5.13 Sequential logic circuit employing D-latches.

When the clock signal is high, the D-latch is transparent so the circuit shown in Fig. 5.13 operates like an asynchronous sequential logic circuit. In order to avoid the state transition hazard problem associated with asynchronous sequential logic circuits, the clock signal must satisfy certain conditions, which we now describe with respect to the timing diagram of the clock signal shown in Fig. 5.14.

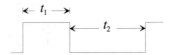

Fig. 5.14 Clock signal for the sequential logic circuit in Fig. 5.13.

In order to guarantee the appropriate operation of a sequential logic circuit that employs D-latches, the clock signal must satisfy the following conditions:

- $t_1 \geq \Delta_{latch}$; Δ_{latch} is the delay of a D-latch. This condition ensures that the next state variable Y has enough time to travel from D to Q so it can be latched correctly.
- $t_1 < \Delta_{C/L(min)}$; $\Delta_{C/L(min)}$ is the shortest delay of the combinational logic circuit. This restriction is a direct result of the latch being transparent during t_1. This condition prevents the potential state transition hazard problem in an asynchronous sequential logic circuit. When this condition is satisfied, the changes caused by the new y still being latched will not come back to the combinational circuit input in the same clock cycle.
- $t_2 \geq \Delta_{C/L(max)}$; $\Delta_{C/L(max)}$ is the longest delay of the combinational circuit. This requirement allows the slowest signal to settle down before the next transition. This ensures the latching of the correct new state variables.
- Input X should not be changed until the latches settle down at their outputs (Q). This can be ensured by changing X during time interval t_2. In this case, t_2 has to be lengthened according to the time of changing X so that this change can propagate completely through the combinational logic circuit. This prevents incorrect state transitions from occurring.

The above requirements are quite restrictive since t_1 must be long enough to correctly latch the new state variable values, but not so long that erroneous values can be stored. In general, it is very difficult to operate a circuit in such a manner properly.

5.4   D-Flip-Flop

The design of a clock signal to be used in a sequential logic circuit can be significantly simplified if we can ensure that the memory element is never (or almost never) allowed to be transparent. This requires the use of an edge-triggered memory element, which responds to the rising or falling edge of the clock signal. The D-flip-flop is such a device. Unlike a level-controlled device, an edge-triggered memory element captures data into storage at the extremely short time intervals when the clock goes through its transitions.

The D-flip-flop shown in Fig. 5.15 is implemented by cascading two D-latches controlled by complementary signals (L and $\overline{L}$). This structure is known as a master-slave structure, in which the master D-latch receives a signal from outside and passes it, under the control of the clock, to the slave D-latch. Since the master and slave D-latches are controlled by complementary signals, the path from D to Q is never allowed to be transparent.[4]

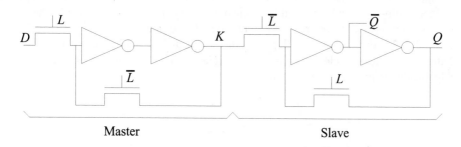

Fig. 5.15 Master-slave structure to form a D-flip-flop.

Fig. 5.16 illustrates the behavior of this D-flip-flop. The latch delay times are ignored in this timing diagram. The output of the master D-latch (K) follows the changes of input D when clock L is high. The slave D-latch accepts K as its input. As noted, since the two D-latches are controlled by complementary signals, K is blocked from entering the slave D-latch while it is changing. By the time when clock L

---

[4] This is only true in an ideal case. In reality, the circuit may be transparent for a very brief time due to the uneven delays of different signals. The designer has to ensure the transparent condition only occurs for a very brief duration of time so that the correct D-flip-flop operation is not compromised. This is usually very easy to achieve.

changes to low, the last signal appearing at K is captured by the master D-latch. Now K becomes the input of the slave D-latch. Note that although the slave D-latch is transparent when clock L is low, its input is held steady by the master D-latch. The overall behavior of the flip-flop is that the falling edge of L triggers it to capture input D at that time into storage.

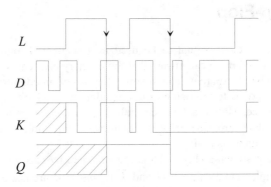

Fig. 5.16 Behavior of a negative-edge-triggered D-flip-flop.

Since the flip-flop in Fig. 5.15 operates in synchronization with the falling edge of signal L, it is called a negative-edge-triggered flip-flop. Similarly, a positive-edge-triggered flip-flop can be formed by exchanging the control signals of the latches so that the master and slave are transparent when $L = 0$ and $L = 1$, respectively. The symbols of both types of flip-flops are illustrated in Fig. 5.17. We use the Greek letter ϕ to indicate a clock signal. The triangle at the control input symbolizes the edge-triggered property of the circuit.

Negative edge-triggered flip-flop   Positive edge-triggered flip-flop

Fig. 5.17 Symbols of flip-flops.

5.5 One-Phase Clock Systems

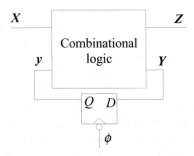

Fig. 5.18 Sequential circuit incorporating D-flip-flops.

The use of D-flip-flops in a sequential circuit is illustrated in Fig. 5.18.[5] We consider the timing diagram shown in Fig. 5.19 to design the clock ϕ used in a sequential circuit incorporating negative edge-triggered D-flip-flops.

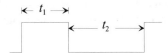

Fig. 5.19 Timing diagram for the design of an edge-triggering clock.

In order to ensure the appropriate operation of the circuit in Fig. 5.18, the clock in Fig. 5.19 must satisfy the following conditions:

- $t_1 \geq \Delta_{latch(master)}$; $\Delta_{latch(master)}$ is the delay of the master D-latch. This allows next state Y to travel from D to K (see Fig. 5.16) so that it can be latched by the master D-latch correctly.

- $t_2 \geq \Delta_{latch(slave)}$; $\Delta_{latch(slave)}$ is the delay of the slave D-latch. This is required to ensure that next state Y that has been latched by the master D-latch can travel from K to Q (see Fig. 5.16). Next state Y can then be latched by the slave D-latch correctly.

- $t_1 + t_2 \geq \Delta_{C/L(max)} + \Delta_{latch(master)} + \Delta_{latch(slave)}$; $\Delta_{C/L(max)}$ is the longest delay time of the combinational logic circuit. This condition allows the signals to settle down before the next transition. When $\Delta_{C/L(max)} \gg (\Delta_{latch(master)} + \Delta_{latch(slave)})$, this requirement can be simplified into $t_1 + t_2 \geq \Delta_{C/L(max)}$.

---

[5] Again, only a single representative D-flip-flop is shown in Fig. 5.18.

- Any changes to input X are to be made after the falling edge of the clock signal ϕ. This ensures that the input and state variable are in synchronization.

The clock for a circuit employing positive-edge-triggered D-flip-flops can be designed similarly by exchanging the restrictions for t_1 and t_2.

The benefit of using edge-triggered flip-flops instead of level-controlled latches in a sequential circuit is that the upper bound restriction of the clock signal (the latch should not be turned on longer than the shortest combinational circuit delay) is removed. This implies that we can always find a valid clock signal for an edge-triggered sequential circuit. All we need to do is to slow the clock down by increasing its cycle time to satisfy the above conditions.

Since in many case, $\Delta_{C/L(max)} \gg (\Delta_{latch(master)} + \Delta_{latch(slave)})$, the length of the clock cycle $(t_1 + t_2)$ is commonly dominated by $\Delta_{C/L(max)}$. If it is desirable to have a duty cycle of 50% for clock ϕ, it can be achieved by setting the transition edge at the middle of a clock cycle so that the clock stays at 1 and 0 for the same amount of time.

The sequential circuits illustrated in Fig. 5.13 and Fig. 5.18 have only one stage of functional unit (i.e., the combinational logic). Fig. 5.20 shows a sequential circuit that has two stages of combinational logic (C/L). Note that each of the D-flip-flops in Fig. 5.20 symbolizes a bank of memory elements, collectively called a register. The registers, controlled with the same clock ϕ, are used to capture associated C/L outputs so that they are available in the next clock cycle. Multiple stages of functional units are commonly used in systems such as pipelines and array processors. We have a detailed discussion on these and other systems later in this book.

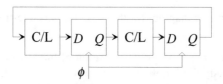

Fig. 5.20 Sequential circuit with two stages of combinational logic.

5.6 Two-Phase, Non-Overlapping Clock Systems

The edge-triggered sequential circuit implementation shown in Fig. 5.20 is straightforward. However, it is sometimes desirable to implement the registers with latches instead of flip-flops, so that 50% of the memory elements, and thus their area, power consumption, and delay, are eliminated. Recall that the difficulty of determining the timing in a latch-based sequential circuit comes from the fact that the latch is transparent when it is turned on by the control signal. When there is more than one C/L stage in the circuit, the problem of transparency can be easily resolved. We can control the latch-based memory elements with a clock signal arrangement that guarantees no continuous loops are formed at any time in the system.

We call such a system a two-phase, non-overlapping clock system. The system employs two identical, but phase-shifted clock signals ϕ_1 and ϕ_2, such as those shown in Fig. 5.21.

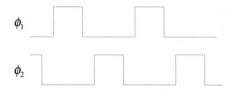

Fig. 5.21 Two-phase, non-overlapping clock signals.

The non-overlapping property of these clock signals ensures that the latches driven by ϕ_1 and ϕ_2 will not be transparent simultaneously. When the ϕ_1-controlled latches are accepting data appearing at their inputs, the ϕ_2-controlled latches are operating in their self-sustaining modes, and vice versa. Theoretically, a clock signal ϕ and its complementary signal $\bar{\phi}$ can be used in a two-phase, non-overlapping clock system. However, the unavoidable delay introduced by the inverter used to generate $\bar{\phi}$ violates the requirement of non-overlapping, as shown in Fig. 5.22.

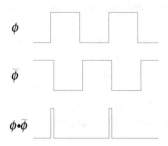

Fig. 5.22 Overlapping between ϕ and $\bar{\phi}$.

In practice, a small amount of overlapping between the clock signals, as long as it is shorter than the shortest delay between two latch stages, can be tolerated. This condition is usually satisfied since the clock overlapping caused by an inverter is often smaller than that of the combinational circuit between two latch stages.

A more serious and challenging problem, called clock-skew, occurs in a two-phase clock system if ϕ_1 and ϕ_2 travel through different paths with different delays. The unequal path delays can bring the two clock signals significantly out of phase.

The difference between the delays of ϕ_1 and ϕ_2 is defined as the amount of clock-skew.

The clock signals in Fig. 5.21 have less than 50% duty cycles, so that an error margin is introduced between the falling edge of ϕ_1 and the rising edge of ϕ_2. A similar error margin is also provided between the falling edge of ϕ_2 and the rising edge of ϕ_1. These error margins must be set to be longer than the maximum amount of clock-skew.

To understand how the two-phase, non-overlapping clock signals are used in a latch-based sequential circuit, consider the output of a latch controlled by clock signal ϕ_1 as shown in Fig. 5.23.

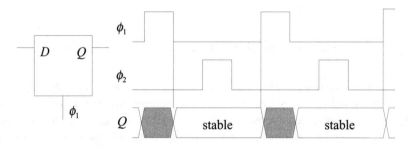

Fig. 5.23 Output Q of a latch controlled by ϕ_1.

The shaded area in the waveform of Q may be unstable since ϕ_1 is high; so the latch is transparent. Shortly after the falling edge of ϕ_1, Q becomes a stable signal because the latch is closed (ϕ_1 is low). Signal Q will remain stable until ϕ_1 rises again. In other words, Q is stable in the time interval when ϕ_1 is low.

The non-overlapping property of the clock signals ensures that the rising and falling of ϕ_2 are complete during the time interval when ϕ_1 is low. Therefore, the output of a latch controlled by ϕ_1 is guaranteed to be stable when a ϕ_2-controlled latch is accepting its input. We call the output of a ϕ_1-controlled latch a stable-ϕ_2 (s-ϕ_2) signal. No glitches can occur if we apply only s-ϕ_2 signals to the inputs of ϕ_2-controlled latches.

Similarly, the output of a ϕ_2-controlled latch is a stable-ϕ_1 (s-ϕ_1) signal. The correct operations of ϕ_1-controlled latches are guaranteed if we apply only s-ϕ_1 signals to their inputs.

In summary, the timing problem of a latch-based sequential circuit is solved by observing the following requirements:

- Only s-ϕ_1 signals can be applied to the D inputs of ϕ_1-controlled latches.
- Only s-ϕ_2 signals can be applied to the D inputs of ϕ_2-controlled latches.
- Different stable signals (s-ϕ_1 and s-ϕ_2 signals) must not be mixed in the same combinational logic stage, which is defined as the logic from one latch stage to the next latch stage.

- The output of a combinational logic stage is of the same type as its inputs (i.e., a combinational logic circuit accepting s-ϕ_1 signals produces s-ϕ_1 signals; a combinational logic circuit accepting s-ϕ_2 signals produces s-ϕ_2 signals).

Fig. 5.24 shows the assignment of clock signals to the latches in a sequential circuit according to these requirements. In some circumstances, additional latches must be added to satisfy the above stable signal requirements.

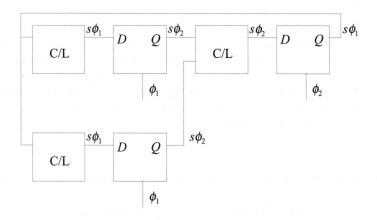

Fig. 5.24 Assignment of clock signals.

The timing of a two-phase, non-overlapping clock system is designed as follows. Consider the timing diagram shown in Fig. 5.25.

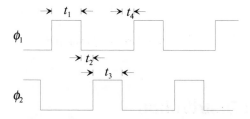

Fig. 5.25 Design of a two-phase, non-overlapping clock system.

The appropriate operation of a latch-based multiple stage sequential circuit can be obtained by satisfying the following timing requirements:

- t_1, $t_3 \geq \Delta_{latch}$. This allows the latches to capture data correctly.
- t_2, $t_4 \geq$ maximum amount of clock-skew. This ensures that the non-overlapping property will not be violated.

- $t_1 + t_2 \geq \Delta_{C/L(max)} + \Delta_{latch}$, $t_3 + t_4 \geq \Delta_{C/L(max)} + \Delta_{latch}$, where $\Delta_{C/L(max)}$ is the maximum delay of a combinational logic stage. This condition allows the signals to settle down at the latch inputs before they are clocked in.

Based on the above timing requirements, the minimum clock cycle time (thus the maximum clock frequency) is determined by the slowest combinational logic stage. It is important to balance the stage delay times so that the fast stages do not sit idle waiting for the slow stages to complete during each clock cycle. A technique called retiming can be applied to relocate the latches for the purpose of balancing stage delays.

An example of retiming is shown in Fig. 5.26. The original circuit is shown in Fig. 5.26(a), in which stage 1 and stage 2 have delays of 100 ns and 50 ns, respectively. The clock cycle time is thus determined to be 100 ns, which implies stage 2 will be idle for half of the clock cycle. The retiming result is shown in Fig. 5.26(b). Stage 1 has been divided into stage 1a and stage 1b, which have delay times of 75 ns and 25 ns, respectively.[6] The latches are relocated between stages 1a and 1b. The clock cycle time after retiming is 75 ns.

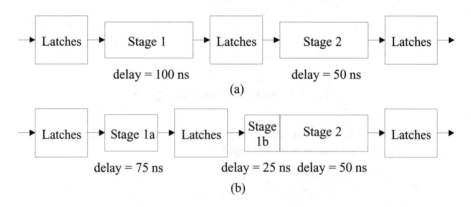

Fig. 5.26 Retiming example.

5.7   Clock Distribution

The distribution of clock signals to different parts of a chip is a major challenge to VLSI designs. Any experienced designer can testify that the routing of clock signals in a high performance circuit is always treated as a "full-custom" operation. In addition to the skew problem we have mentioned above, other important issues in the routing of clock signals are cross-talks and load balancing. A routing scheme called the H-tree scheme and its variations can be used to distribute the clock signal

---

[6] This is an ideal case to simplify the explanation.

to different parts of a chip so that every part has approximately the same distance from the clock source. The H-tree routing scheme is illustrated in Fig. 5.27.

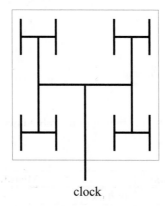

clock

Fig. 5.27 H-tree clock routing scheme.

5.8   Sequential Circuit Design

We conclude this chapter by going through a simple example to review the general design process of sequential circuits. Readers who need further information about sequential circuit design should consult one of the digital logic design books listed at the end of the chapter.

Example 5.1

In this example, we design a sequential logic circuit that controls and maneuvers a mobile robot lawn mower (Fig. 5.28). The robot is equipped with a light sensor and an obstacle sensor. The light sensor detects the ambient light intensity. The obstacle detects obstacles in the path of the robot. The outputs of the sensors are shown in Fig. 5.29.

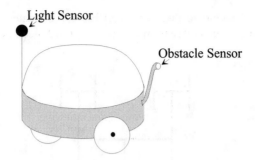

Fig. 5.28 Illustration of the robot lawn mower.

Sensor	Signal	Meaning	
Light	x	0: dark	1: bright
Obstacle	y	0: obstacle free	1: obstacle detected

Fig. 5.29 The sensor signals and their meanings.

The sequential circuit uses two output signals (p and q) to control two robot motors, respectively. Each motor drives a wheel. A motor turns on if its control signal is 1 and turns off if the control signal is 0. The robot thus moves according to Fig. 5.30.

pq	Left motor	Right motor	Robot
00	off	off	stops
01	off	on	turns left
10	on	off	turns right
11	on	on	goes straight

Fig. 5.30 Relations between control signals and robot movements.

The maneuver of the robot around obstacles is carried out using the following algorithm:
- In order to avoid accidents, the robot stops moving when there is not enough ambient light ($x = 0$). It resumes its movement when the lighting is adequate ($x = 1$).[7]
- The robot always goes straight ($pq = 11$) initially after it resumes its movement.

---

[7]This feature is for the benefit of the humans around the robot.

- The first obstacle in the robot's path ($y = 1$) after it resumes its movement causes the robot to turn left (i.e., $pq = 01$). The robot continues to turn until no obstacle is detected.
- The robot remembers its previous turning direction. When a new obstacle is detected, the robot turns in a direction different from its previous one. For example, if the robot has turned right last time to avoid an obstacle, it turns left until no obstacle is detected.

A state diagram is created in Fig. 5.31 to graphically capture the above requirements for the sequential circuit. The state diagram is created as a Mealy machine to reduce the number of states (see Problem 5.10 for a Moore machine implementation). Note that the state assignment (i.e., assigning a binary code to a state) can significantly affect the performance (circuit complexity, power consumption, speed, etc.) of a finite state machine. We will not discuss state assignment in this book, but readers are encouraged to review the related issues.

The circles in the state diagram indicate states, each of which is labeled with $s_1 s_0$, a two-bit state variable. There are four states. The arrows represent possible state transitions, each of which is labeled with xy/pq, where xy is the sensor condition that causes the transition and pq is the required motor control signal. For example, if the robot is currently in state 00 and $xy = 0$- (i.e., $x = 0$ and y is a *don't care*). The machine remains in state 00 and produces 00 on its output.

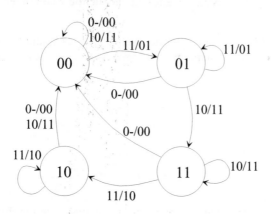

Fig. 5.31 Mealy machine state diagram for the mobile robot controller.

The meanings of individual states are listed in Fig. 5.32.

State	Meaning
00	Next obstacle will be avoided by turning left
01	Turning left
11	Next obstacle will be avoided by turning right
10	Turning right

Fig. 5.32 States of the mobile robot controller.

The state diagram is then converted into the state table shown in Fig. 5.33.

$s_1 s_0$	$xy = 0-$ $S_1 S_0 p q$	$xy = 10$ $S_1 S_0 p q$	$xy = 11$ $S_1 S_0 p q$
00	0000	0011	0101
01	0000	1111	0101
11	0000	1111	1010
10	0000	0011	1010

Fig. 5.33 State table for the mobile robot controller.

The optimized Boolean equations for next state variables S_1 and S_0 and outputs p and q are created using the Kanaugh maps in Fig. 5.34. The result Boolean equations are shown in (5.6).

$$
\begin{aligned}
S_1 &= s_1 xy + s_0 x \overline{y}, \\
S_0 &= \overline{s_1} xy + s_0 x \overline{y}, \\
p &= s_1 x + x \overline{y}, \\
q &= \overline{s_1} x + x \overline{y}.
\end{aligned}
\tag{5.6}
$$

The block diagram for this mobile robot sequential circuit is shown in Fig. 5.35. Two memory elements (e.g., D-flip-flops) are used to store the state variable. The combinational circuit (C/L) implements the Boolean equations given in (5.6). It is desirable to provide a reset function (not shown) so that the sequential circuit can be brought into a known state as needed (e.g., at power-up). Any of the CMOS logic circuit structures (e.g., complementary logic) described in Chapter 2 can be used. The technique explained in Section 5.6 can be applied to determine the maximum clock frequency (see Problems 5.9 and 5.10). A means must also be provided to reset the robot, either manually or automatically at the power-up time.

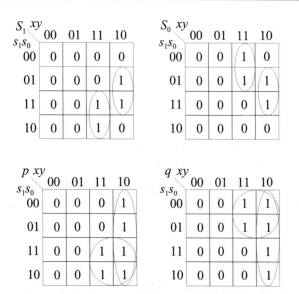

Fig. 5.34 Knaugh maps for mobile robot controller design.

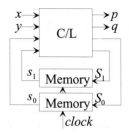

Fig. 5.35 Block diagram for the mobile robot controller circuit.

5.9   Summary

In this chapter, we have introduced two basic CMOS memory elements, the D-latch and the D-flip-flop. The D-latch is a level-controlled device; so it is transparent when its control signal is asserted. In contrast, the D-flip-flop is an edge-triggered device, which captures an input signal into storage in response to the edge of a clock signal. The use of both types of memory elements to construct sequential circuits has been demonstrated. A number of approaches to use clock signals to synchronize the operations of a sequential circuit have been described.

5.10 To Probe Further

Sequential Circuit Design:

- J. F. Wakerly, *Digital Design Principles and Practices*, *2nd* *Edition*, Prentice Hall, 1994.

- V. P. Nelson, H. T. Nagle, B. D. Carroll, and J. D. Irwin, *Digital Logic Circuit Analysis & Design*, Prentice-Hall, 1995.

- G. De Micheli, R. K. Brayton, and A. Sangiovanni-Vincentelli, "Optimal state assignment for finite state machines," *IEEE Transactions on CAD/ICASI*, CAD-4, 1985, pp. 269-285.

Two-Phase Clocking:

- D. Noice, R. Mathews, and J. Newkirk, "A clocking discipline for two-phase digital systems," *Proceedings, International Conf. on Circuits and Computers*," IEEE Computer Society, 1982, pp. 108-111.

Clock Distribution:

- E. G. Friedman, ed., *Clock Distribution Networks in VLSI Circuits and Systems*, IEEE Press, 1995.

Retiming:

- C. E. Leiserson, F. M. Rose, and J. B. Saxe, "Optimizing synchronous circuitry by retiming," *Proc. Third Caltech Conf. on VLSI*, Randal Bryant, ed., Computer Science Press, Rockville, MD, 1983, pp. 87-116.

5.11 Problems

5.1   Setup time and hold time are two important parameters of a clocked circuit. Setup time and hold time specify the minimum amounts of time that the input should be held stable before and after the triggering clock transition, respectively. What is the setup time and hold time of the D-flip-flop in Fig. 5.15?

5.2   Explain why is it possible to have a negative setup time, which implies that the input does not have to be ready until after the triggering clock transition.

5.3   From the viewpoint of power consumption, explain why it is important to have reasonable fast rise time and fall time.

5.4   Verify that for the circuit in Fig. 5.5, a voltage increment at v_A causes oscillation; assume that $v_A = v_B$ initially.

5.5   Modify the D-flip-flop shown in Fig. 5.15 to include an asynchronous reset function. This allows the sequential circuit to be brought into a known initial state. Show a transistor level circuit diagram.

5.6   Modify the D-flip-flop shown in Fig. 5.15 into a JK-flip-flop.

5.7   Follow the guidelines described in this chapter to design a 2-phase non-overlapping clock to be used in the sequential circuit of Fig. 5.26b. Use the latch shown in Fig. 5.10. The circuit should be operated in its highest possible speed.

5.8   The block diagram shown in Fig. 5.36 suggests a way to implement a 64-bit adder. Only a single one-bit full adder and a number of registers are used. Show the state diagram, state and output equations for this 64-bit adder. Design a 2-phase non-overlapping clock for this circuit and provide your answer with a timing diagram. Assume all transistors are of the same size. Consider only gate capacitance, effective channel resistance, and output diffusion capacitance. All other parasitic elements are ignored. How long does it take for your circuit to produce the sum?

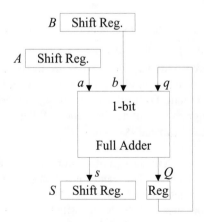

Fig. 5.36 64-bit adder.

5.9   Use complementary logic to implement the combinational logic portion of Fig. 5.35. Assume D-flip-flops are used and determine the maximum clock frequency at which the sequential circuit operates correctly. Provide a reset function.

5.10   Redesign the mobile robot lawn mower controller described in Example 5.1 as a Moore machine. Determine the maximum clock frequency at which your design operates correctly. Provide a reset function.

5.11   Design the control circuitry for a garage door opener. A single button controls the opening and closing of the door. Pressing the button opens a closed door or closes an opened door. For safety reasons, the direction of a moving door should be reversed immediately if the button is pressed.

5.12   Design a combinational lock with three buttons, 1, 2, and R. The numerical buttons are used to enter a code and the R button is used to reset the lock. To open the lock, a user would first push the reset button R and then enter a code of 3 digits. If a mistake is made in the process, the R button can be used to start over again. Your circuit should produce an output of 0 to open the lock after the correct code has been entered. Show the design process and the schematic circuit diagram of this lock for the code 1-2-1.

5.13   Use a minimum number of inverters and nMOS transistors to design a CMOS positive-edge-triggered "toggle" flip-flop with an asynchronous reset (R). A toggle flip-flop is a storage device with a control input T. If $T = 1$ when the clock ticks, the state of the flip-flop changes; otherwise, it remains the same as the previous state.

5.14   A sequential circuit has the following state and output equations:
       $Y_1 = \overline{X_1} X_2 y_1 + \overline{X_2}\ \overline{y_1} y_2$, $Y_2 = X_2 \overline{y_1} + \overline{X_1} X_2 y_1 + \overline{X_2}\ \overline{y_1} y_2$, $Z = \overline{X_2}\ \overline{y_1} y_2$.
       y_1, y_2 are the current state variables. Y_1, Y_2 are the next state variables. X_1, X_2 are inputs. Z is the output. Implement this circuit with CMOS complementary logic circuits and static D-flip-flops. The design objective is to operate this circuit at the highest operating frequency. What is this frequency? Use transistors with $L = 1\ \mu m$ and $W = 4\ \mu m$.

5.15   This problem deals with the design of a control unit for a simple coin-operated candy vending machine. The candy costs 20¢, and the machine accepts nickels (5¢) and dimes (10¢). Change should be returned if more than 20¢ is deposited. No more than 25¢ can be deposited on a single purchase; therefore, the maximum change is one nickel. A block diagram of the candy machine and a state diagram for its control unit are given in Fig. 5.37 and Fig. 5.38, respectively.

The control unit has two inputs, N and D, which are outputs of the coin detector. The coin detector generates a 1 on signal N if a nickel is deposited and a 1 on signal D if a dime is deposited. The N and D lines automatically reset to 0 on the next clock pulse. We shall assume that it is physically impossible to insert two coins at the same time, and therefore we cannot have $N = D = 1$ in the same clock period. The control unit has two outputs, R and C. The candy is released by a 1 appearing on signal R, and a nickel in change is released by a 1 appearing on signal C. The states of the control unit represent the total amount of money deposited for the current purchase.
Design this candy machine control unit.

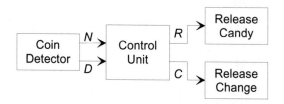

Fig. 5.37 Block diagram of a candy machine.

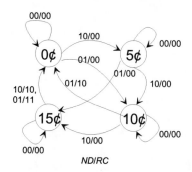

Fig. 5.38 State diagram for a candy machine.

Chapter 6 Alternative Logic Structures

Out of many, one ...

The essentials of CMOS design and analysis have been covered in the previous five chapters. A non-exhaustive list of CMOS structures has been shown in Chapter 2 (see Fig. 2.16). Complementary logic circuits and pass-transistor/transmission-gate logic circuits have been introduced and used in the design examples. This chapter presents several alternative CMOS logic structures, which include open-drain circuits, pseudo-nMOS circuits, dynamic circuits, and BiCMOS circuits.

The availability of more than one way to implement a function offers a design with extra dimensions of flexibility. Different logic structures can be used together in the design of a system. We begin our discussion with a summary of complementary logic circuits and pass-transistor/transmission-gate logic circuits, which we establish as the baseline CMOS circuit structures. Complementary logic and pass-transistor/transmission-gate logic are extensively used in standard cell libraries because of their robustness.

6.1 Complementary Logic Circuits

Complementary logic circuits are named after the fact that each of them contains two complementary transistor switch networks. The pull-up network, when turned on, connects a 1 (V_{DD}) to the gate output. The pull-down network, when turned on, supplies a 0 (V_{SS}) to the gate output.

The design of complementary logic circuits has been discussed in Chapter 2. Complementary CMOS logic circuits have a number of properties:

- Easy to design.
- Simple to analyze.
- Robust operation.
- Variable supply voltage.
- Static circuit.
- Ratioless logic.
- Inverting logic gates (e.g., NAND, NOR, etc.) only.[1]
- $2n$ transistors (n nMOS and n pMOS) for an n-input gate.
- Input capacitance of 2 transistor gates for each input for an n-input gate.
- Maximum logic swing (V_{DD} to V_{SS}).
- Almost no static power dissipation.
- Power dissipation only during logic transition.

---

[1] A non-inverting logic gate can be formed by applying inverters at either the inputs or the output of a complementary logic gate.

Note that the above summary is a broad overview and exceptions to each point do exist. It should only be used as a rule-of-thumb insight list by a designer to make intelligent decisions.

6.2   Pass-Transistor/Transmission-Gate Logic

Pass-transistor/transmission-gate logic circuits provide a design with another dimension of freedom by allowing input signals in addition to V_{DD} and V_{SS} to be steered to their outputs. A summary of pass-transistor logic properties is provided as follows.
- Inverting or non-inverting logic.
- Clipping logic (i.e., reduced logic swing).
- An nMOS pass-transistor produces weak-1 signals.
- A pMOS pass-transistor produces weak-0 signals.
- A pass-transistor should not be used to drive another pass-transistor.
- Potential floating hazards if not designed correctly.
- A 2^n-to-1 multiplexer requires $(2^{n+1} - 2)$ pass-transistors plus n inverters.

The weak signal problem can be resolved by replacing the pass-transistors with t-gates (transmission-gates). Fig. 6.1 shows the effective on-resistance of a t-gate.

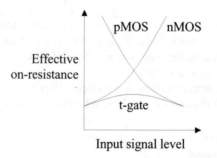

Fig. 6.1   Effective on-resistance of a t-gate.

The properties of transmission-gate logic are listed below:
- Non-clipping switches.
- Twice as many transistors as an equivalent pass-transistor circuit.
- Complementary signals needed to control transmission-gates.
- Higher input capacitance than a pass-transistor counterpart.
- Higher power consumption (due to the extra inverters and higher input capacitance).

6.3 Pseudo-nMOS Logic

Removing one of the switch networks in a complementary logic circuit results in an open-drain logic circuit. Usually the pull-up network is eliminated since pMOS transistors are inherently slower than their nMOS counterparts. Fig. 6.2 shows an example of an open-drain inverter, which is created by removing the pMOS transistor (i.e., the pull-up network) of a complementary inverter.

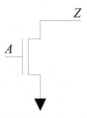

Fig. 6.2 An open-drain inverter.

It is easy to verify that the circuit in Fig. 6.2 only implements half of the inverter function. When $A = 1$, the transistor turns on and pulls the output Z down to V_{SS} (0). The transistor is turned off when $A = 0$ so the output is floating. Due to the capacitance load (not shown) at Z, the output retains its value for a time determined by the small reverse-bias p-n junction current and the load capacitance value.

In order to correctly operate this open-drain logic circuit, a pull-up resistor must be added between output Z and V_{DD} to avoid the floating condition when $A = 0$. This structure is illustrated in Fig. 6.3. When the pull-down transistor is off, R_{pu} is responsible for the charging of the capacitance at Z so that it can be brought up to 1. According to the RC timing model, the pull-up time is estimated by the product of R_{pu} and the capacitance value at the output Z.

In Fig. 6.3, when $A = 1$, a voltage-divider is formed by the pull-up resistor R_{pu} and the nMOS transistor. The result is that output voltage v_Z depends on the ratio between the pull-up resistance R_{pu} and the effective channel resistance of the nMOS transistor R_n:

$$v_Z = \frac{V_{DD} R_n}{R_{pu} + R_n}, \text{ when } A = 1. \tag{6.1}$$

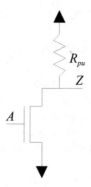

Fig. 6.3 Open drain inverter with a pull-up resistor.

From (6.1), it can be seen that v_Z cannot reach 0 unless $R_n = 0$. In theory, the value of v_Z can be brought closer to 0 by increasing the value of R_{pu}. However, in practice we would keep the value of R_{pu} as low as possible to minimize its area and improve the pull-up time of the circuit. The following analysis provides an approach to determine the smallest value of R_{pu} that produces an acceptable 0 at output Z.

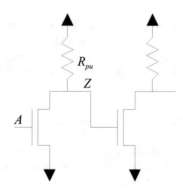

Fig. 6.4 Open-drain inverter driving another open-drain gate.

Assume that the output of an open-drain circuit is used to drive another open-drain stage, as shown in Fig. 6.4. When the driver stage produces a 0 at its output Z (i.e., $A = 1$), the necessary condition for v_Z to be considered a 0 by the second stage is that its value should be below the threshold voltage V_{tn} of the nMOS transistor. Assume that $V_{tn} = 0.2\,V_{DD}$, we need

$$v_Z = \frac{V_{DD}R_n}{R_{pu} + R_n} < 0.2V_{DD}. \tag{6.2}$$

The condition of (6.2) can be simplified and rewritten into

$$R_{pu}/R_n > 4 \qquad (6.3)$$

We call R_{pu}/R_n the pull-up to pull-down ratio. Typically we select this ratio to be a value between 5 and 10.

In the CMOS technology, the pull-up resistor is implemented with a pMOS transistor that is permanently turned on by having its gate connected to V_{SS}. An example of an open-drain inverter using a pMOS transistor as its pull-up resistor is shown in Fig. 6.5. This structure is similar to the inverter in the now obsolete nMOS technology and is thus called a pseudo-nMOS inverter.[2]

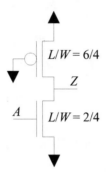

Fig. 6.5 Pseudo-nMOS inverter.

Considering the fact that the effective channel resistivity of the pMOS transistor is about 2.5 times that of the nMOS transistor, the pull-up to pull-down ratio is satisfied by the following transistor dimension relationship:

$$\frac{L_{pu}}{W_{pu}} = 3\frac{L_{pd}}{W_{pd}}, \qquad (6.4)$$

in which L_{pu} and W_{pu} are the effective channel length and channel width of the pull-up transistor, respectively, and L_{pd} and W_{pd} are the effective channel length and width of the pull-down network, respectively.

The transistor dimensions for a pseudo-nMOS inverter are given in Fig. 6.5. Fig. 6.6 shows the transistor dimensions for a pseudo-nMOS 2 input NAND gate. The NAND gate has two nMOS transistors connected in series in the pull-down network. The effective pull-down channel length is thus twice that of a single transistor $(L_{pd}/W_{pd} = 4/4)$. The condition of (6.4) requires the pull-up pMOS channel to be

---

[2] NMOS logic uses a depletion mode transistor with a negative threshold voltage as a pull-up resistor.

proportionally lengthened (L_{pu}/W_{pu} = 12/4) to satisfy the pull-up to pull-down ratio requirement.

Fig. 6.7 shows the configuration of a pseudo-nMOS 2 input NOR gate. In the case of a NOR gate, since the nMOS transistors are connected in parallel, the channel dimensions of the pMOS transistor remain the same as in an inverter (L_{pu}/W_{pu} = 6/4). This result comes from the worst case consideration in which only one nMOS transistor is turned on. If both nMOS transistors are turned on, the overall pull-down resistance will be reduced so the pull-up to pull-down ratio would be further improved. For this reason, a NOR gate is often preferred over a NAND gate in pseudo-nMOS logic circuits.

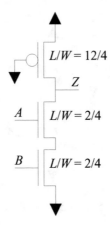

Fig. 6.6 Transistor dimensions of pseudo-nMOS NAND gate.

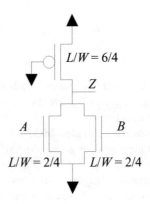

Fig. 6.7 Transistor dimensions of pseudo-nMOS NOR gate.

A pseudo-nMOS logic gate of n inputs generally has $(n + 1)$ transistors, which is approximately a 50% reduction from the $2n$ transistors in a functionally identical complementary logic circuit. However, the use of transistor counts to evaluate the circuit areas can be misleading in this case since the pull-up transistor is larger (sometimes significantly) than the transistors used in a complementary logic circuit.

The pseudo-nMOS logic has a reduced logic swing due to the voltage dividing structure when the pull-down network is turned on. Another disadvantage is that the pseudo-nMOS logic has non-zero static power dissipation if it outputs a 0. When the pull-down network is turned on, a current path between V_{DD} and V_{SS} is formed so power is dissipated although the circuit is not switching.

One unique property of pseudo-nMOS logic is its capability of forming hardwired logic by connecting the outputs of two or more logic circuits together. For example, hardwiring the outputs of two open-drain inverters and providing an appropriate pull-up pMOS transistor combines the inverters into a NOR gate. This is equivalent to forming an AND function between the inverter outputs. Hardwired logic is not allowed in complementary logic since connecting outputs with unequal signals create a fighting situation.

Fig. 6.8 illustrates an example application of this hardwiring capability of pseudo-nMOS logic circuits. In this example, multiple service requesters are connected to the same request line of a service provider. This type of circuit can be used in the connection of peripheral devices to a mother board. Each service requester connects to the request line through an open-drain stage. The request line is normally high because of the pull-up pMOS transistor. The service provider monitors the request line through an inverter.

A low signal on the request line indicates that one or more service requester has turned on its output transistor that pulls down the request line. A priority scheme such as daisy chaining (not shown) can then be used to identify the service requester or to resolve the situation when two or more service requesters have pulled down the request line simultaneously.

The advantage of this architecture is apparently in the fact that the operation is independent of the number of service requesters connected to the request line. The adding or removing of service requesters does not affect the operation. Notice that all service requester output stages are connected in parallel, which allows the use of a fixed size pMOS pull-up transistor.

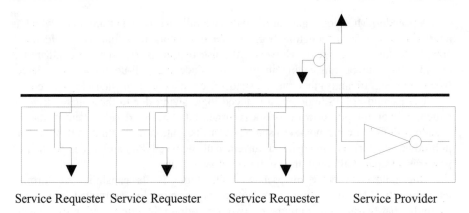

Service Requester Service Requester   Service Requester   Service Provider

Fig. 6.8 Request line connecting multiple service requesters and a service provider.

A summary of pseudo-nMOS logic circuit properties is provided here:
- Ratioed logic.
- Static power dissipation when output = 0.
- $(n + 1)$ transistors for n-input gates.
- Reduced logic swing.
- Input capacitance of 1 transistor gate for each input of an n-input gate.
- Hard-wired logic possible.

6.4   Programmable Logic Array

One popular application of pseudo-nMOS logic structure is the implementation of programmable logic arrays, which are programmable two-level logic circuits. Programmable logic array (PLA) was a term originally created to describe a class of standalone devices that allow users to program their functionalities. The capability of these user programmable devices were quite limited and have been mostly replaced by the significantly more powerful field programmable gate arrays (FPGAs). The structure of a field programmable gate array will be discussed in Chapter 8.

In the context of VLSI design, the designers are not interested in using the PLA structure to provide the user with programmability. Rather, we would like to develop the PLA as a highly regular, multiple output logic structure for the purpose of automatic layout generation.

Fig. 6.9 shows the block diagram of a PLA. It consists of two levels of combinational logic functions and one level of input buffers. The input buffer provides both inverting and non-inverting buffers for the inputs. The two levels of combinational logic are used together to implement sum-of-products logic functions. The block marked as "AND" is called the and-plane, which is responsible for the generation of all product terms needed to form the logic functions. The block marked as "OR" is called the or-plane which OR's selected product terms together to form the desired logic functions.

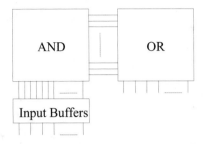

Fig. 6.9 Block diagram of a PLA.

The pseudo-nMOS logic structure is chosen for its simplicity (only one switching network to program) to implement PLAs in the CMOS technology. Both the and-plane and or-plane are pseudo-nMOS NOR logic structures. A pseudo-nMOS NOR structure has the advantage of a constant size pull-up. Since all the pull-down transistors are connected in parallel, the pull-up size does not have to be modified when the number of pull-down transistors is changed. Also, from the layout point of view, it is physically easier to add or remove transistors when they are all connected in parallel. The following example demonstrates the formation of a PLA.

Example 6.1

The following functions are to be implemented with a pseudo-nMOS PLA:

$$Z_1 = A\overline{B} + \overline{A}B$$
$$Z_2 = AB$$
$$Z_3 = A + B$$
$$Z_4 = AB + \overline{A}\,\overline{B}$$

The complexity of a PLA is determined by the numbers of its inputs, its product terms, and its outputs. This example has 2 inputs (A and B) and 4 outputs (Z_1, Z_2, Z_3, and Z_4). The product terms from the and-plane are shared in the or-plane. A product term (e.g., AB), although it is needed in more than one output, has to be generated only once.

Normally a multiple function minimization should be performed in the design of a PLA. The goal is to minimize the number of unique product terms instead of individual output functions. There are 6 unique product terms in this example. A PLA implementation of this example is shown in Fig. 6.10.

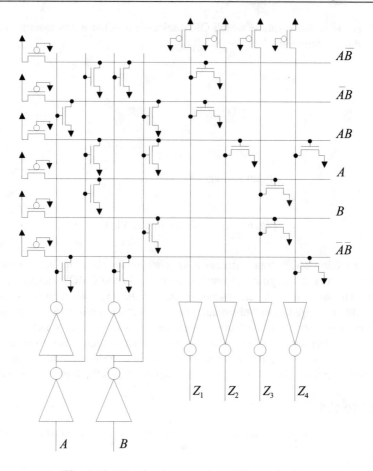

Fig. 6.10 PLA implementation of Example 6.1.

The pull-up to pull-down ratio required by pseudo-nMOS circuits must be followed in the design of a PLA. Notice that De Morgan's Theorem has been applied in the formation of the and-plane so that the required products are formed with NOR functions. Due to the regularity of the PLA structure, its layout can be generated automatically by assigning nMOS transistors to appropriate locations in the array according to the given logic functions.

A simple analysis can show that the PLA structure does not utilize the silicon area efficiently. For example, the transistor density of the and-plane is always at or below 50% since a variable and its complement would never appear in a product term at the same time. A technique called PLA folding has been proposed to improve the transistor density. For example, depending on the functions being implemented, PLA folding looks for a way to physically place two inputs into a column in the and-plane. PLA folding also manages to put two product terms in one row in the or-plane. Two outputs may also be folded into one column. Several references to PLA folding techniques are given at the end of this chapter.

6.5   Dynamic CMOS Logic

The nodes of a CMOS circuit are commonly modeled as capacitors. The capacitance values of them depend on the circuit geometry and, in the case of an output node, the number of fan-outs. We called a node that is electrically isolated from the rest of the circuit a floating node or a high impedance node.

Assume that a node with capacitance C_n has an initial voltage of V_0 before it becomes a floating node. The charge stored in C_n is $Q_n = C_n V_0$. In an ideal case when there is no leakage current, the amount of charge and thus the voltage will be held indefinitely. In practice, a CMOS circuit has a small leakage current so C_n will eventually be discharged. While the charge accumulated in C_n is being depleted, the voltage on C_n changes accordingly.

The extremely low leakage current of a CMOS circuit allows a floating node to retain a large portion of its charge for a period of time. Dynamic CMOS circuits rely on this property to store signals on circuit nodes. Since the charge leaks away, the storage nodes in a dynamic circuit need to be periodically refreshed. Refreshing is typically done in synchronization with a clock signal. Since the storage nodes must be refreshed before their values deteriorate, the clock should be running at or above a minimum frequency.

In contrast, a static circuit does not have a minimum operating frequency requirement.[3] The stable output value of a static circuit is retained indefinitely as long as the power is on and the inputs are not changed. The clock in a static circuit can be slowed down or completely stopped. This is often desirable when a circuit is being debugged since we can step through its state transitions, one clock cycle at a time.

Before we discuss and analyze dynamic CMOS logic structures, we use the following example to investigate the relationship between the node capacitance and its minimum refreshing frequency.

Example 6.2

A CMOS circuit node has a capacitance of 0.01 pF. How long can a signal value be held at this node? Calculate the minimum refreshing frequency of this circuit.

Assume the node is initially charged up to $V_{DD} = 2.5$ V. The amount of charge stored on the capacitance is $Q_1 = CV_1 = 2.5 \times 10^{-14}$ coulomb. If we can tolerate at most a 0.5 V voltage drop, then the allowable amount of lost charge is $Q_2 = CV_2 = 5 \times 10^{-15}$ coulomb. If the leakage current is 0.1 nA, then it will take

---

[3] All the circuits that we have discussed so far are static circuits.

$$t = \frac{5 \times 10^{-15}}{0.1 \times 10^{-9}} \text{s} = 50 \ \mu s \tag{6.5}$$

to cause a 0.5 V voltage drop at this node.

Consider that in every clock cycle, the capacitance will be loaded with a new value. The minimum operating frequency of this circuit is then 20 MHz. Many CMOS circuits operate at much higher clock frequencies (hundreds of MHz) so no specific refreshing arrangement will be required.

Fig. 6.11 shows the basic structure of a dynamic CMOS logic gate. Similar to the pseudo-nMOS logic circuit, a dynamic logic circuit eliminates one of the switch networks from a complementary logic circuit. Usually the pMOS switch network is eliminated, but there are exceptions. We will discuss this issue in more detail.

Two complementary transistor switches (Q_p and Q_e) are connected in series with the pull-down network. A clock signal ϕ is used to synchronize the operation of the dynamic logic structure. When ϕ is 0, transistor Q_p is turned on so the output node Z modeled by capacitance C_Z is charged toward V_{DD}. Note that C_Z denotes the total capacitance at node Z. We call this operation step the precharging phase. The clock ϕ must stay at 0 long enough so that $v_Z = V_{DD}$ at the end of the precharging phase. Since transistor Q_e is off during the entire precharging phase, no conducting current path between V_{DD} and V_{SS} exists.

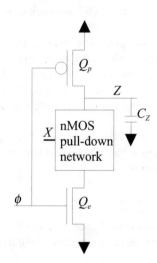

Fig. 6.11 Basic dynamic CMOS logic circuit structure.

The precharging phase ends when ϕ turns 1. Transistor Q_p is turned off and transistor Q_e is turned on. The circuit then operates in its evaluation phase to determine its output. Depending on the structure of the nMOS pull-down network and input X, output Z either temporarily retains its value of 1 (V_{DD}) or discharges to 0 through the pull-down network and transistor Q_e. We further analyze the behavior of a dynamic circuit by considering the dynamic 2-input NAND gate shown in Fig. 6.12.

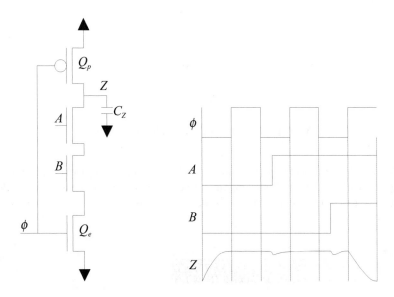

Fig. 6.12 Waveform diagram of a dynamic 2-input NAND gate.

Fig. 6.12 shows the relationship between the clock ϕ, inputs A and B, and output Z of a dynamic 2-input NAND gate. Consider the first clock cycle. When ϕ is 0, output Z is precharged to 1. When ϕ turns high, the circuit goes into its evaluation phase. In the evaluation phase, Q_e is turned on. Since both inputs A and B are low, the pull-down network is off. The output Z goes into a high impedance state and holds the precharged value of 1, which is the expected result of a NAND gate.

In the second clock cycle, input A changes to 1 in the middle of the precharge phase. This change causes a small dip in the waveform of output Z. The voltage dip is caused by a situation called charge sharing. We will discuss charge sharing in more detail. During the evaluation phase, the pull-down network is still off; so output Z is 1.

The third clock cycle illustrates a case where the output would be pulled down to 0. During the precharging phase, signal B turns high, which causes another small dip in output Z. In the evaluation phase, Q_e turns on to complete the pull-down current path and output Z drops down to 0. Note that the output of a dynamic logic circuit is only valid after the evaluation is completed.

The small dips occur when a pull-down transistor turns on during a precharging phase caused by a situation called charge sharing. Consider the circuit in Fig. 6.13.

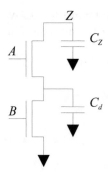

Fig. 6.13 Circuit illustrating charge sharing.

Assume the following initial condition: $A = B = 0$, C_d is discharged and C_Z is charged up to V_{DD}. If now A is changed to 1 so C_z and C_d are connected in parallel, a portion of the charge stored in C_Z is distributed to C_d. The voltage at node Z becomes:

$$v_Z = \frac{V_{DD}C_Z}{C_Z + C_d} \tag{6.6}$$

Depending on the capacitance value of C_d, the voltage drop may or may not affect the integrity of the signal at Z. For example, if $C_d = C_Z$, then v_Z is changed from V_{DD} into $0.5V_{DD}$. In a static circuit, charge sharing voltage dips are not important, since the lost charge will be replenished and the signal restored. This is, however, not the case in a dynamic circuit when the value holding node is at a high impedance state. Any charge lost by charge sharing during the evaluation phase is lost and will not be replenished until the next precharging phase.

The charge sharing problem can be resolved by ensuring that all input changes are made within the precharging phase so that not only the output node, but also the internal nodes of the pull-down network will be charged up.

A dynamic logic gate has $(n + 2)$ transistors, where n is the number of inputs. The advantage of a dynamic logic circuit over its pseudo nMOS logic counterpart is that the control transistors do not have to satisfy the pull-up to pull-down ratio requirement. Most importantly, the power saving advantage of CMOS circuits is restored since the complementary switches Q_p and Q_e prevent a continuous current path to form between V_{DD} and V_{SS}.

While dynamic logic circuits have a number of benefits, they are haunted by a serious limitation. Dynamic logic circuits cannot be cascaded. In other words, a dynamic logic circuit cannot be used to drive another dynamic logic circuit. To appreciate the cause of this problem and its solution, consider the circuit in Fig. 6.14 in which a dynamic inverter is driven by another dynamic inverter.

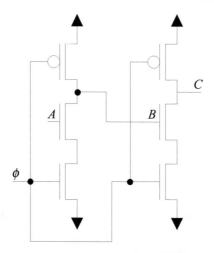

Fig. 6.14 Cascaded dynamic inverters.

The waveform of this circuit in Fig. 6.14 is shown in Fig. 6.15. During the first clock cycle, both nodes B and C are first precharged to V_{DD} in the precharging phase. In the evaluation phase, since $A = 0$, B retains its charge so it is equal to 1. This turns on the pull-down network of the second dynamic inverter. Along with the evaluation transistor, the pull-down network of the second dynamic inverter completes the pull-down path. Node C is discharged to logic 0. This result matches with the expected one according to the function of this circuit.

A problem occurs after A changes to 1 in the precharging interval of the second clock cycle. During the evaluation, the pull-down transistor of the first inverter stage is turned on so node B begins to discharge. This operation itself is correct since B is the complement of A. However, the discharging of node B cannot occur instantaneously. Until B drops down to 0 (or at least below the threshold of an nMOS transistor), the second inverter stage sees a 1 at its input. The second stage pull-down network is thus turned on by B, which is still incorrect at this time. The voltage at C thus drops as illustrated.

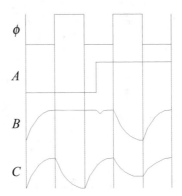

Fig. 6.15 Waveform of two cascading dynamic inverters.

When the discharging of B is completed at a later time, the correct value at B turns off the pull-down network of the second inverter so the discharging stops. However, the charge stored at node C is now partially gone and cannot be replenished. The value at output C is thus corrupted. This problem is especially serious if the interconnection between the two inverters has a long delay so that clock ϕ makes its transition long before the correct B signal arrives at the input of the second stage.

A number of solutions have been proposed to solve the problem of cascading dynamic stages. All of them are based on the principle of postponing the evaluation of a following stage until its previous stage has completed its evaluation. These circuits are called domino logic circuits.

Before we move on to the discussion of domino logic, we summarize the properties of dynamic logic circuits:

- An n-input logic gate requires $(n + 2)$ transistors.
- Minimum clock frequency required.
- Input capacitance of 1 transistor gate for each input of an n-input gate.
- Propagation delay includes precharging time.
- Power dissipation even if the output remains low in multiple clock cycles.
- Cannot be cascaded.

6.6 Domino Logic

The simplest domino solution to the cascading problem of dynamic logic circuits is to add a static inverter at the output of a dynamic logic stage. Such a domino circuit is shown in Fig. 6.16. The static inverter ensures that the output of the circuit is 0 at the end of the precharging phase, which will not incorrectly turn on the pull-down transistors in the following stage. Only if the output of the dynamic circuit is evaluated to a 0 will the static inverter produce a 1 to turn on the pull-down transistors of the next stage. In other words, the following stage will not start its evaluation until the driver stage has completed its evaluation. The drawback of this type of domino logic circuit is that it is only limited to non-inverting logic. Non-inverting logic is not complete.[4]

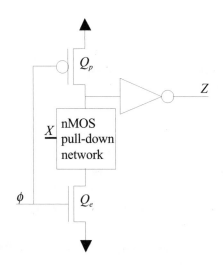

Fig. 6.16 Non-inverting domino logic circuit.

Earlier in this section we have mentioned the possibility of using a pMOS pull-up network to implement a dynamic logic circuit. This turns out to be a key to achieve the domino effect. Fig. 6.17 shows the use of both pull-down based (nMOS) and pull-up based (pMOS) dynamic stages to create a domino logic circuit (zipper logic). This structure does not need a static inverter attached at a dynamic gate output. Instead, the cascaded dynamic logic circuits are organized in a special manner. A dynamic logic circuit consisting of an nMOS pull-down network is used to drive a dynamic circuit implemented with a pMOS pull-up network. A dynamic logic circuit consisting of a pMOS pull-up network is used to drive a dynamic circuit implemented with an nMOS pull-down network. Also notice that the clock signal for the stages implemented with pMOS pull-up networks is the complement of the clock signal for the stages implemented with nMOS pull-down networks.

When $\phi = 0$, the outputs of the first and third stages (with pull-down networks) are precharged to 1 and the output of the second stage (with a pull-up network) is precharged to 0. The precharged signal at node A keeps the pMOS transistors in the second stage off during the precharging phase. In contrast, node B is precharged to 0 which keeps the nMOS transistors in the third stage off. Only after a stage has completed its evaluation can a following stage begin to evaluate.

---

[4] A complete logic is one that can be used to implement any logic functions. For example, a 2-input NAND gate is complete since any logic functions can be implemented by an appropriate number of NAND gates.

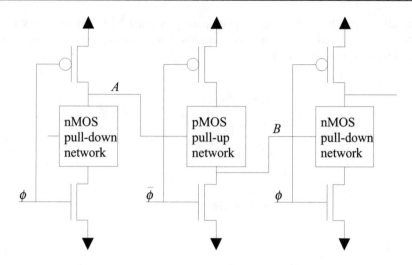

Fig. 6.17 Inverting domino logic circuits.

Other domino structures are also possible as long as the principle of orderly evaluation is ensured.

6.7   Dynamic Memory Elements

In Chapter 5 we developed the basic memory elements for CMOS sequential circuits, the D-latch and the D-flip-flop. In practice, we consider the D-latch to be a static memory element since the stored data can be held indefinitely. However, this static storage is actually achieved through an internal refreshing feedback path; so the CMOS D-latch is sometimes called a pseudo-static D-latch.

The property of a dynamic circuit that uses fewer transistors than its static counterpart can be extended to the design of memory elements. Fig. 6.18 shows the structure of a dynamic D-latch, which has only half of the components of a pseudo-static D-latch. The data are stored at the input capacitance of the inverter.

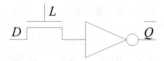

Fig. 6.18 Dynamic D-latch.

Two dynamic D-latches can be cascaded into a master-slave type edge-triggered D-flip-flop. Fig. 6.19 shows a negative edge triggered D-flip-flop. When dynamic D-latches or D-flip-flops are used in a sequential circuit, the minimum operating frequency of the circuit has to be determined in addition to the maximum operating frequency. The minimum operating frequency is determined by the storage node with the smallest capacitance value. The maximum operating frequency is constrained by the combinational logic circuit with the longest propagation delay. Dynamic circuits are often harder to analyze and more difficult to test. Another problem with dynamic circuits is that they are susceptible to radiation. The energy of radiation may cause unexpected leakage current to flow, which can upset dynamic circuits.

Fig. 6.19 Negative edge triggered D-flip-flop.

6.8   BiCMOS Logic Circuits

In this section, we provide a brief introduction to the BiCMOS technology, which combines bipolar and CMOS circuits on the same chip. The objective is to provide the high current-driving capability of bipolar transistors to CMOS circuits. CMOS, although a technology with all the desirable logic circuit properties such as low power, high input impedance, and large logic swing, has a limited current-driving capability.

The large transconductance of a bipolar transistor renders it capable of large output currents and thus higher speeds. Emitter-coupled logic (ECL), a BJT-based technology, is two to five times faster than CMOS at the expense of high power dissipation. BiCMOS technology attempts to provide the best characteristics of both worlds.

Fig. 6.20 shows the implementation of a BiCMOS inverter. The circuit operates as follows. When v_i is low, nMOS Q_2 is turned off. BJT Q_4 has no base current so it is also turned off. On the other hand, pMOS Q_1 turns on and supplies BJT Q_3 with base current, thus turning it on. BJT Q_3 provides a large current to charge the load capacitance and pulls v_o up. When v_i is high, Q_1 and Q_3 turn off, nMOS Q_2 and BJT Q_4 turn on. A large output current quickly discharges the load capacitance and pulls v_o down.

Without resistor R_1, v_o cannot reach V_{DD} since Q_3 turns off when v_o reaches a value of $V_{DD} - 0.7$ V since a BJT requires a forward bias of about 0.7 V across its base-emitter junction. The incorporation of R_1 provides a pull-up path to pull the output node up to V_{DD} through Q_1. A similar function is provided by the incorporation of resistor R_2. Resistors in BiCMOS circuits are implemented with nMOS transistors.

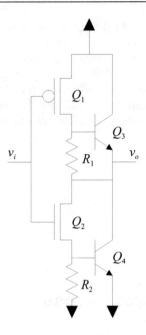

Fig. 6.20 BiCMOS inverter.

The structure shown in Fig. 6.20 can be generalized to convert a complementary CMOS logic circuit into a BiCMOS circuit. A BiCMOS two-input NAND gate is shown in Fig. 6.21. The logic (i.e., pull-up and pull-down) is performed by the CMOS part of the gate. The bipolar part of the gate functions as the output stage.

6.9   Summary

The proper operation of a complementary logic circuit is virtually guaranteed. However, alternative logic structures allow a designer to optimize a design according to specific criteria. After reviewing the properties of complementary logic circuits and pass-transistor/transmission-gate logic circuit, a number of alternative CMOS structures were introduced in this chapter.

The first logic circuit is the open-drain logic circuit which requires a pull-up resistor of appropriate value to operate. A pseudo-nMOS logic circuit uses a pMOS transistor to implement the pull-up resistor. The structure of a programmable logic array (PLA) was introduced as an application of the pseudo-nMOS logic circuit. Dynamic logic circuits and domino logic circuits were introduced. The structures of dynamic latches and flip-flops were discussed. A brief introduction of BiCMOS logic circuits was provided. Rule-of-thumb type features of different CMOS structures were given.

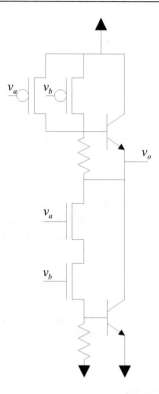

Fig. 6.21 BiCMOS 2-input NAND gate.

6.10 To Probe Further

CMOS Logic Structures:

- V. G. Oklobdzija, *High-Performance System Design: Circuits and Logic*, IEEE Press, 1999.

PLA Minimization:

- J. F. Hill and G. R. Peterson, *Introduction to Switching Theory and Logical Design*, 3rd *Edition*, Wiley, 1981.

- D. D. Givone, *Introduction to Switching Circuit Theory*, McGraw-Hill, 1970.

- M. R. Dagenais, V. K. Agarwal, and N. C. Rumin, "McBOOLE: a new procedure for exact logic minimization," *IEEE Trans. on Computer-Aided Design*, vol. CAD-5, no. 1, Jan. 1986, pp. 229-238.

- B. Gurunath and N. N. Biswas, "An algorithm for multiple output minimization," *IEEE Trans. on Computer-Aided Design*, vol. CAD-8, no. 9, September 1989, pp. 1007-1013.

- E. J. McCluskey, *Logic Design Principles*, Prentice-Hall, 1986.

PLA Folding:

- G. D. Micheli and A. L. Sangiovanni-Vincentelli, "PLEASURE: a computer program for simple/multiple constrained/unconstrained folding of programmable logic arrays," *Proc. 20th Design Automation Conference*, June 1983, pp. 530-537.

Domino Circuits:

- R. H. Krambeck, C. M. Lee, and H. F. S. Law, "High-Speed Compact Circuits with CMOS," *IEEE Journal of Solid-State Circuits*, SC-17(3), June, 1982, pp. 614-619.

BiCMOS Logic Circuits:

- S. Sedra and K. C. Smith, *Microelectronic Circuits*, 4th Edition, Oxford University Press, 1998.

6.11 Problems

6.1 Design a domino logic circuit for $Z = A + BC$.

6.2 Design a pseudo-nMOS circuit for $Z = A \oplus B$. Indicate the dimensions of the transistors.

6.3 Suggest a way to eliminate Q_e, the evaluation transistor, in a dynamic circuit.

6.4 Develop a two-phase clock system to be used in cascaded dynamic logic circuits. Explain how the two-phase clock system can ensure a correct operation.

6.5 Suggest a way to modify the PLA structure so that some of the outputs can be fed back into the inputs to form a synchronous sequential circuit. Minimize the number of additional components used in this modification.

6.6 Use the PLA structure developed in Problem 6.5 to implement a sequential circuit, of which the state diagram is shown in Fig. 6.22.

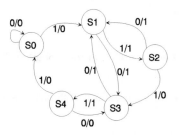

Fig. 6.22 State diagram for problem 6.6.

6.7 Implement the following logic functions with a minimized PLA:
$$y_1 = \overline{x}_2 x_4 + x_1 \overline{x}_2 x_4 + x_1 x_2 \overline{x}_3 \overline{x}_4$$
$$y_2 = \overline{x}_2 x_4 + x_1 x_2 \overline{x}_3 \overline{x}_4 + \overline{x}_3 \overline{x}_4$$
$$y_3 = x_1 \overline{x}_2 x_4 + x_1 x_2 \overline{x}_3 \overline{x}_4 + \overline{x}_3 \overline{x}_4$$

6.8 Design a BiCMOS circuit for a 2-input NOR gate. Use a circuit simulator to verify your design.

6.9 Design a BiCMOS circuit for $Z = \overline{A + BC}$. Use a circuit simulator to verify your design.

6.10 Design a multiplier to multiply two 2-bit numbers and produce a 4-bit product using a minimized PLA. Show a transistor level diagram for your PLA.

6.11 What is the logic function $Z(a, b)$ implemented by the circuit given in Fig. 6.23?

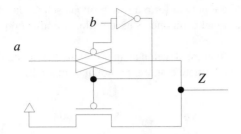

Fig. 6.23 Circuit for Problem 6.11.

6.12   Create a truth table for the pseudo-nMOS circuit shown in Fig. 6.24.

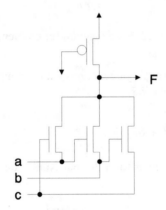

Fig. 6.24 Circuit diagram for Problem 6.12.

6.13 Develop the structure of a dynamic PLA. Explain its operation.

Chapter 7 Sub-System Design

Foundation ...

Programmable logic array (PLA) was introduced in Chapter 6 as a general building block to implement arbitrary logic functions. In this chapter, we continue to discuss the design and use of building blocks beyond basic logic gates. These building blocks include adders, full adder trees, multipliers, read-only memory (ROM), and random-access memory (RAM). In most cases, a required building block can be implemented in more than one way. Our discussion concentrates on structures that are suitable for CMOS implementations.

The choice of a building block implementation often makes a significant difference in the performance of an overall system design. Sometimes these structurally different, functionally equivalent implementations are well documented, such as in the case of standard-cells and IP cores (intellectual property). The selection of a building block among them is then somewhat simplified since their properties are fully known. In other situations such as in a full-custom design, however, the designer often depends only on analytical results to choose an appropriate structure.

The following objectives for creating VLSI structures have been advocated by researchers since the beginning of the VLSI era:

- A VLSI structure should be divided into a number of building blocks, each of which can be designed and analyzed independently of the rest of the circuits. This architectural property, which is referred to as modularity, enables the divide-and-conquer design methodology.

- A preferable VLSI structure consists of only a few types of building blocks, which are replicated and assembled in a regular pattern. This property is called the regularity. Only a few building blocks need to be created and evaluated in the design of a regular architecture. This guideline is based on the fact that silicon is cheap and design activity is expensive.

- A well-designed VLSI structure should avoid the use of long communication paths since they cause long delay times. They also present challenges to the routing procedure. This is called the guideline of local communication. In an ideal case, nearest neighbor interconnections should dominate the communication pattern.

A tessellation (tiling) technique can be used to lay out a VLSI system with high degrees of modularity, regularity, and local communication. The building blocks designed for tessellation are created to have built-in nearest neighbor interconnections. Routing is automatically achieved by abutting building blocks together.

The advantage of this type of layout structure is that very little routing external to the building blocks is needed. A tessellated design also allows a VLSI structure to be readily expanded to accommodate problems with larger operands (e.g., 32 bits vs.

16 bits). Automatic layout generators called tilers have been created for these structures.

Many interesting and powerful VLSI architectures, such as systolic and wavefront arrays, have been developed with these objectives in mind for many important applications (see Chapter 12).

7.1 Adders

Ripple-Carry Adder

Addition is a common arithmetic function that is covered by virtually every book written for digital logic design. The basic element available for building an adder is a full adder. Fig. 7.1 shows that a one-bit full adder accepts two 1-bit operands (a_i, b_i) and one carry-in bit (c_{i-1}) and produces a result consisting of 1 sum bit (s_i) and 1 carry-out bit (c_i). An alternative view of the one-bit full adder is that it accepts three 1-bit operands and produces a two-bit sum.

Fig. 7.1 One-bit full-adder.

A number of one-bit full adders can be cascaded to form a ripple-carry adder (see Fig. 7.2). The ripple-carry adder is extremely modular since it contains only one-bit full adders. The structure is also very regular. All full adders in the ripple-carry adder are connected to their neighbors in the same way. The interconnections within the ripple-carry adder are limited to between neighboring full adders.

The full adder cell can be laid out to accept a carry-in bit (c_{i-1}) from its right edge and produces a carry-out bit (c_i) at its left edge. This layout will facilitate the use of tessellation to the design of a ripple-carry adder. Fig. 7.3 illustrates that when two full adder cells are placed side by side, their power rails (V_{DD} and V_{SS}) as well as their carry paths are connected. It is also quite simple to expand a ripple-carry adder. An n-bit ripple-carry adder can be expanded to accept ($n + m$)-bit operands by incorporating m additional full-adder cells.

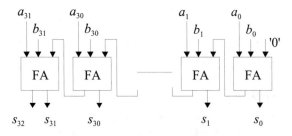

Fig. 7.2 Ripple carry adder.

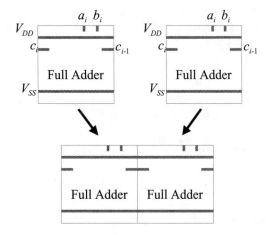

Fig. 7.3 Full adder tessellation.

A ripple-carry adder has a computation time that is proportional to the number of bits in its operands. The wider the operands, the longer does it take the ripple-carry adder to complete its addition. Whether this presents a limitation or not on the use of a ripple-carry adder naturally depends on the application on hand. A designer must be able to balance the benefits and limitations of a specific structure so that an intelligent building block selection can be made.

General expressions can be developed for the purpose of building block evaluation and comparison. For example, the area (A_{rca}), computation time (Δ_{rca}), and power consumption (P_{rca}) of an n-bit ripple-carry adder can be estimated as

$$A_{rca} = nA_{FA} \tag{7.1}$$

$$\Delta_{rca} = n\Delta_{FA} \tag{7.2}$$

$$P_{rca} = nP_{FA} \qquad\qquad (7.3)$$

where A_{FA}, P_{FA}, and Δ_{FA} are the area, computation time, and power consumption of the one-bit full adder being used, respectively.

Carry Select Adder

The addition time of a ripple-carry adder can be improved with a modified structure called the carry select adder. It retains the layout regularity of a ripple-carry adder. The principle of a carry select adder is to use one ripple-carry adder to execute an addition assuming that the carry-in is 1. Another ripple-carry adder is used to execute the same addition assuming that the carry-in is 0. The real carry-in computed in a previous stage is used to select one of the two sums with a multiplexer. Fig. 7.4 shows an example of an 8-bit carry select adder with a 4-4 staging. The 4-4 staging specifies that two stages are to be used, each of which performs the addition of 4 bits.

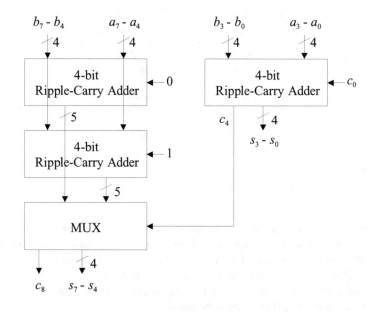

Fig. 7.4 8-bit carry select adder.

All three 4-bit ripple-carry adders in the carry select adder work in parallel. The addition time is thus the delay of the first stage 4-bit ripple-carry adder plus the delay of a multiplexer. For longer operands, the use of stages with identical lengths does not necessarily lead to an optimal delay. This is caused by the different delays of the selecting carry bits which have to go through different levels of multiplexers. A circuit level analysis can be performed to determine the best staging scheme for a given technology (see Problem 7.1).

Carry-Lookahead Adder

The biggest limitation of a ripple-carry adder is in its long computation time since the carry bit has to be propagated from one full adder to another. Many digital design textbooks describe the use of a carry-lookahead operation to remove the propagation of the carry.

The carry-out bit c_i at the ith full adder stage is calculated as

$$c_i = a_i b_i + (a_i + b_i) c_{i-1} \qquad (7.4)$$

There are two terms in (7.4). The first term $a_i b_i$ is called the carry generation function, which decides whether a carry-out bit is generated at the ith stage. The OR operation in the second term $(a_i + b_i)c_{i-1}$ is called the carry propagation function, which determines if a carry-in bit (c_{i-1}) generated at a previous stage can propagate through the ith stage to the next one. Applying (7.4) recursively we can express the carry-out bits as functions of the operand bits only:

$$
\begin{aligned}
c_0 &= a_0 b_0 \\
c_1 &= a_1 b_1 + (a_1 + b_1) a_0 b_0 \\
    &= a_1 b_1 + a_1 a_0 b_0 + a_0 b_1 b_0 \\
c_2 &= a_2 b_2 + (a_2 + b_2)(a_1 b_1 + a_1 a_0 b_0 + a_0 b_1 b_0) \\
    &= a_2 b_2 + a_2 a_1 b_1 + a_2 a_1 a_0 b_0 + a_2 a_0 b_0 b_1 + a_1 b_2 b_1 + a_1 a_0 b_2 b_0 + a_0 b_2 b_1 b_0
\end{aligned}
\qquad (7.5)
$$

......

According to (7.5), the carry-in bit of every stage can be generated independently of the previous stages. A carry-lookahead adder applies this principle to produce a sum. Theoretically the computation time of a carry-lookahead adder is independent of the propagation of the carry value. In reality the carry-lookahead circuit becomes excessively complicated for wide operands.

Assuming that we would like to maintain a fan-in of less than 10 for a CMOS logic gate, the implementation of a carry-lookahead adder would run into some significant limitations. It is observed in (7.5) that an n-bit carry-lookahead adder requires logic gates which has as many as $(n + 1)$ inputs. If we assume that the maximum fan-in of a CMOS logic gate is 10, a carry-lookahead adder is then limited to 8-bit operands.[1] A longer adder can be constructed out of smaller carry-lookahead adders with the carry rippling through them.

---

[1] While the fan-in limitation can be circumvented by implementing a multiple level logic circuit, the propagation delay goes up with the number of levels in the circuit.

Conditional Sum Adders

Various adder structures have been developed to utilize the switch-like characteristic of MOSFETs. The conditional sum adder is such a structure that uses pass-transistor logic to achieve an addition time significantly faster than a ripple-carry adder.

Consider that we need to add a_i and b_i, which are the ith bits of operands A and B, respectively. Recall that this operation cannot be completed until c_{i-1}, the carry-out of the $(i - 1)^{th}$ position, is available. In the conditional sum algorithm, instead of waiting for the arrival of the carry value, conditional sum and conditional carry are generated by considering both possible values of the carry-in bit. In the following expressions, the result of adding a_i, b_i, and either a 0 or a 1 carry-in bit is a two bit number ($C_i^0 S_i^0$ and $C_i^1 S_i^1$).

$$C_i^0 S_i^0 = a_i + b_i + 0$$
$$C_i^1 S_i^1 = a_i + b_i + 1$$

$$(7.6)$$

where the superscript of a conditional result bit indicates the condition that it was generated. Example 7.1 demonstrates the generation of these conditional results.

Example 7.1

Generate the conditional sum bits and conditional carry bits for adding $A = (1101101)_2$ and $B = (0110110)_2$.

The results are shown below. Notice that all conditional sums and conditional carries can be generated in parallel.

	bit 6	bit 5	bit 4	bit 3	bit 2	bit 1	bit 0
$A =$	1	1	0	1	1	0	1
$B =$	0	1	1	0	1	1	0
S_i^0	1	0	1	1	0	1	1
C_i^0	0	1	0	0	1	0	0
S_i^1	0	1	0	0	1	0	0
C_i^1	1	1	1	1	1	1	1

Fig. 7.5 Conditional sums and conditional carries.

The sum of A and B can be found by using the ith carry-in bit to select the correct conditional bits in the $(i + 1)^{th}$ bit position. This is demonstrated in Fig. 7.6.

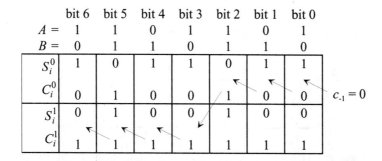

Fig. 7.6 Addition with conditional sums and conditional carries.

In Fig. 7.6, we assume that the carry-in bit going into bit 0 (c_{-1}) is 0; so S_0^0 and C_0^0 are the correct results and they are selected. $C_0^0 = 0$, which determines that the correct sum bit and carry bit at bit 1 are S_1^0 and C_1^0, respectively. The arrows indicate the selection procedure. The sum of A and B is determined to be $(10100011)_2$.

There is no benefit in performing the conditional sum addition as shown in Example 7.1 since it is essentially the same as a ripple carry addition with extra overhead to create the conditional sum bits and carry bits. This is going to be changed with a multiple level conditional sum addition. We illustrate this approach in the next example.

Example 7.2

Consider two bits at a time and generate conditional sums and conditional carries for the addition in Example 7.1.

The results are shown in Fig. 7.7. The grouping of bits began at the most significant bit (bit 6). Each box contains two conditional sum bits and one conditional carry bit. Notice that since the operands have 7 bits, bit 0 is grouped by itself; so a one-bit box was created for it.

Fig. 7.8 shows how the 2-bit conditional sum and carry table is used to perform the addition. Again, the arrows indicate the selection of conditional sums and conditional carries. The sum of A and B is determined in 4 selections, which is only one half of what was needed in Fig. 7.6 when only one bit position was considered at a time.

	bit 6	bit 5	bit 4	bit 3	bit 2	bit 1	bit 0
$A =$	1	1	0	1	1	0	1
$B =$	0	1	1	0	1	1	0
S_i^0	0	0	1	1	0	1	1
C_i^0	1		0		1		0
S_i^1	0	1	0	0	1	0	0
C_i^1	1		1		1		1

Fig. 7.7 Two-bit conditional sums and conditional carries.

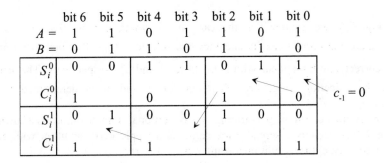

	bit 6	bit 5	bit 4	bit 3	bit 2	bit 1	bit 0	
$A =$	1	1	0	1	1	0	1	
$B =$	0	1	1	0	1	1	0	
S_i^0	0	0	1	1	0	1	1	
C_i^0	1		0		1		0	$c_{-1} = 0$
S_i^1	0	1	0	0	1	0	0	
C_i^1	1		1		1		1	

Fig. 7.8 Addition with 2-bit conditional sums and conditional carries.

We now introduce an efficient way to generate the two-bit conditional sum and carry values. Consider the table in Fig. 7.9, which is a combination of Fig. 7.5 and Fig. 7.7.

In Fig. 7.9, we have added another index (in parentheses) to indicate the level of bit grouping. The number of bits grouped together is equal to 2^j at level j. The conditional sums and conditional carries at level j are labeled as $S_i^0(j)$ and $C_i^0(j)$, respectively. The second level ($j = 1$) of conditional sums and carries is generated with the values created in the first level ($j = 0$).

	bit 6	bit 5	bit 4	bit 3	bit 2	bit 1	bit 0
$A =$	1	1	0	1	1	0	1
$B =$	0	1	1	0	1	1	0
$S_i^0(0)$	1	0	1	1	0	1	1
$C_i^0(0)$	0	1	0	0	1	0	0
$S_i^1(0)$	0	1	0	0	1	0	0
$C_i^1(0)$	1	1	1	1	1	1	1
$S_i^0(1)$	0	0	1	1	0	1	
$C_i^0(1)$	1		0		1		
$S_i^1(1)$	0	1	0	0	1	0	
$C_i^1(1)$	1		1		1		

Fig. 7.9 Parallel determination of 2-bit conditional sums and conditional carries.

Fig. 7.9 shows how $S_i^0(1)$ and $C_i^0(1)$ are generated by considering the conditional sums and carries in the first level. Take bits 6 and 5 as an example. Since the assumed condition is $C_5 = 0$, $S_5^0(0)$ and $C_5^0(0)$ are selected. $C_5^0(0) = 1$ so $S_6^1(0)$ and $C_6^1(0)$ are selected, which, along with $S_5^0(0)$, form $S_6^0(1), S_5^0(1)$ and $C_6^0(1)$. All three groups of conditional sums and conditional carries at level two can be generated concurrently.

Generalizing the principles that we have discussed above, we create the table in Fig. 7.10 to complete the conditional sum addition algorithm.

The conditional sums and carries at level 3 ($j = 2$) are generated with the values at level 2 ($j = 1$). Fig. 7.10 also shows how this three-level conditional sum addition is used to find the sum of A and B. Now the number of selections in the critical propagation delay path has been reduced to 3. The implementation of this three-level carry-select adder is left as an exercise (see Problem 7.3). In Example 7.3 we present an example of a two-level carry-select adder.

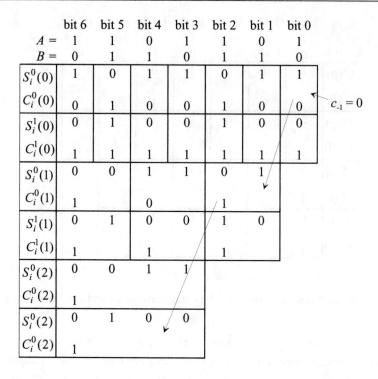

Fig. 7.10 Three levels of conditional sums and conditional carries.

Example 7.3

Develop a schematic diagram for a three-bit conditional sum adder.

The design is presented in Fig. 7.11. A two-level conditional sum adder is needed to add two three-bit operands. Two types of building blocks are needed. The conditional cells (CC) generate conditional sum bits and conditional carries at level 1. The 2-to-1 multiplexers provide the carry controlled selections.

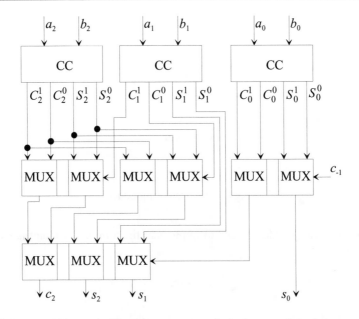

Fig. 7.11 Two-level three-bit conditional sum adder.

7.2 Full Adder Tree

Assume that we would like to design a circuit to add n numbers ($n > 2$) together. One possible approach to this design problem is to set up an adder along with an accumulating register. If we initialize the accumulating register with the first number and then add the other $(n-1)$ numbers to it, the final sum will be ready in $(n-1)$ clock cycles. Let us look at an example.

Example 7.4

Design an adder to compute the sum of 4 four-bit numbers.

The first thing we have to determine is the length of the accumulating register so that it will not overflow. Assume that the operands are unsigned numbers. The largest value of a four-bit unsigned number is $(15)_{10}$. Four of them will give a maximum sum of $(60)_{10}$. The number of bits needed for the accumulating register is thus $\lceil \log_2 60 \rceil = 6$ bits. This adder is shown in Fig. 7.12. A 6-bit adder and a 6-bit register are used to form a circulating adder. To simplify the operation, the register is provided with a "Clear" control line to initialize its value to "000000." The four operands are then added sequentially to the register.

input operands

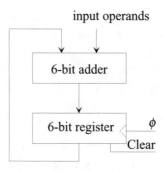

Fig. 7.12 Circulating adder.

If a ripple carry adder is used to implement the 6-bit adder then the expression for delay time is:

$$\Delta_{adder} = 4 \times (6\Delta_{FA} + \Delta_{reg}), \tag{7.7}$$

where Δ_{FA} is the delay time of a one-bit full adder and Δ_{reg} is the delay time of the register.

Now we present a different approach called the full adder tree to add up these four numbers. The full adder tree structure is a combinational circuit. The only type of functional block that we are going to use is the 1-bit full adder.

The numbers to be added are labeled as:

$A = a_3 a_2 a_1 a_0;$
$B = b_3 b_2 b_1 b_0;$
$C = c_3 c_2 c_1 c_0;$
$D = d_3 d_2 d_1 d_0.$

Fig. 7.13 shows the structure of a full-adder tree that is used to add up these four numbers by adding bits in the same column position together. The propagation delay of the full adder tree is determined to be 6 Δ_{FA} by identifying its critical path which is marked by the bold-face arrows in Fig. 7.13.

We must point out that the circulating adder shown in Fig. 7.12 has one advantage over the full adder tree illustrated in Fig. 7.13. It has the capability of latching the result in the register. The full adder tree is a combinational circuit. It is an interesting problem to determine general expressions for estimating the area and delay time of a full adder tree that is used to add n m-bit numbers (see Problem 7.4).

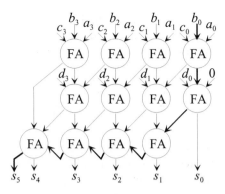

Fig. 7.13 Full adder tree.

7.3   Parallel Multipliers

High speed multiplication is another critical function in a range of VLSI applications. In this section we discuss a number of VLSI oriented parallel multipliers. These multipliers may be used when the conventional add-and-shift multiplication (a sequential operation) cannot meet the design requirement.

Braun Array Multipliers

First we review the procedure of finding the product of $A = a_{m-1}...a_2a_1a_0$ and $B = b_{n-1}...b_2b_1b_0$. The decimal values of A and B can be expressed as

$$A_v = \sum_{i=0}^{m-1} a_i 2^i \tag{7.8}$$

$$B_v = \sum_{j=0}^{n-1} b_j 2^j \tag{7.9}$$

The product P of A and B has $(m + n)$ bits. Its value is

$$P_v = A_v \times B_v = \left(\sum_{i=0}^{m-1} a_i 2^i \right) \times \left(\sum_{j=0}^{n-1} b_j 2^j \right) = \sum_{i=0}^{m-1} \sum_{j=0}^{n-1} (a_i \times b_j) 2^i 2^j \tag{7.10}$$

In the binary number system, the multiplication of a multiplicand bit a_i by a multiplier bit b_j can be implemented by the logic AND function to produce a partial product bit $a_i b_j$. The above expression can thus be rewritten as

$$P_v = \sum_{i=0}^{m-1} \sum_{j=0}^{n-1} a_i b_j 2^{i+j} = \sum_{k=0}^{m+n-1} p_k 2^k \tag{7.11}$$

Realizing that the multiplication is simply the addition of the $a_i b_j$'s ($0 \leq i \leq m-1$, $0 \leq j \leq n-1$), a full adder tree can be built to calculate the product bits p_k's ($0 \leq k \leq m+n-1$). The result is a Braun array multiplier.

Example 7.5

Build a Braun array multiplier to perform the multiplication of $A = a_3 a_2 a_1 a_0$ and $B = b_3 b_2 b_1 b_0$.

The expressions for the product bits are:

$p_0 = a_0 b_0$
$p_1 = a_1 b_0 + a_0 b_1 + c$
$p_2 = a_2 b_1 + a_1 b_1 + a_0 b_2 + c$
$p_3 = a_3 b_0 + a_2 b_1 + a_1 b_2 + a_0 b_3 + c$
$p_4 = a_3 b_1 + a_2 b_2 + a_1 b_3 + c$
$p_5 = a_3 b_2 + a_2 b_3 + c$
$p_6 = a_3 b_3 + c$
$p_7 = c$

where c is the carry from a previous column.

Fig. 7.14 shows a full adder tree created to perform the multiplication. We refer to this type of multiplier as the Braun array multiplier. In the layout of a Braun array multiplier, we would like to arrange the full adder cells so that a rectangular array can be formed.

Notice the modularity and regularity of this Braun array multiplier. Also, except for the distribution of the input signals, all interconnections are limited to nearest neighbors. Instead of forming the partial product bits ($a_i b_j$) outside of the array multiplier and feeding them in, AND functions are incorporated into the full adder cells. Notice that all full adder cells in the same row use the same b_j. Also, all full adder cells in the same carry path use the same a_i. This property allows a mesh of signal lines to be formed between FA cells. Horizontal signal lines carry b_j's and vertical signal lines carry a_i's. At each intersection of a horizontal signal line and a vertical signal line, the a_i signal and the b_i signal are ANDed for the full adder cell.

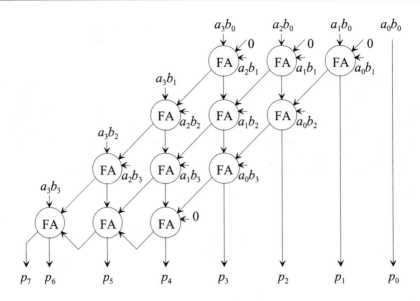

Fig. 7.14 Braun array multiplier.

The Braun array multiplier shown in Fig. 7.14 has the following properties:

$$A_{bam} = 16A_{AND} + 12A_{FA}$$
$$\Delta_{bam} = \Delta_{AND} + 6\Delta_{FA}$$
(7.12)

In (7.12), A_{bam} and Δ_{bam} are the area and delay time of the 4-bit by 4-bit Braun array multiplier, respectively. A_{AND} and Δ_{AND} are the area and delay time of a two-input AND gate, respectively. A_{FA} and Δ_{FA} are the area and delay time of a full adder, respectively. The area and delay time expressions can be generalized for an m-bit by n-bit Braun array multiplier (see Problem 7.6).

The Braun array multiplier performs unsigned multiplication. Many applications require signed multiplication. Before we describe an array multiplier that performs direct two's complement multiplication, we look at a way that allows the Braun array to do signed multiplication. In fact, this method is valid for any unsigned multipliers.

A two's complement number (positive or negative) $X = x_{n-1}x_{n-2}...x_1x_0$ has the value of

$$X_v = -x_{n-1}2^{n-1} + \sum_{i=0}^{n-1} x_i 2^i$$
(7.13)

The product of $A = a_{n-1}...a_1a_0$ and $B = b_{n-1}...b_1b_0$ can be written as

$$a_{n-1}b_{n-1}2^{2n-2} - a_{n-1}\sum_{j=0}^{n-2}b_j 2^{n+j-1} - b_{n-1}\sum_{i=0}^{n-2}a_i 2^{n+i-1} + \sum_{i=0}^{n-2}\sum_{j=0}^{n-2}a_i b_j 2^{i+j} \qquad (7.14)$$

The operations described in (7.14) can be performed in an unsigned array multiplier provided that all partial products which involve a sign bit (a_{n-1} or b_{n-1}) and a nonsign bit are complemented. An adjustment is carried out to the final result by adding a correction term $2^{2n-1} + 2^n$. This adjustment comes from the operation of obtaining the two's complement of a given number and is fixed for given word lengths (see Problem 7.7).

The multiplication of $A = 1101$ (-3) and $B = 0101$ (5) is demonstrated in Fig. 7.15.

```
         1101  = -3
      X) 0101  =  5
         0101
         1000
         0101
         0111
      01100001
   +) 10010000   adjustment
      11110001  = -15
```

Fig. 7.15 Two's complement multiplication with an unsigned multiplier.

Baugh-Wooley Array Multiplier

The Baugh-Wooley array multiplier was developed to perform two's complement multiplication. This approach does not need any pre-treatment or post-treatment to be carried out. The multiplication of two numbers A and B was given in (7.14). The negative terms in the expression can be transformed to avoid the use of subtractors. This is done by observing the following relations:

$$\overline{a}_i = (1 - a_i); \text{ so } -a_i = -1 + \overline{a}_i$$

$$-a_{n-1}\sum_{j=0}^{n-2}b_j 2^{n+j-1} = 2^{2n-1} + 2^{2n-2} + \overline{a}_{n-1}2^{2n-2} + a_{n-1}2^{n-1} + a_{n-1}\sum_{j=0}^{n-2}\overline{b}_j 2^{m+j-1} \qquad (7.15)$$

$$-b_{n-1}\sum_{i=0}^{n-2}a_i 2^{n+i-1} = 2^{2n-1} + 2^{2n-2} + \overline{b}_{n-1}2^{2n-2} + b_{n-1}2^{n-1} + b_{n-1}\sum_{i=0}^{n-2}\overline{a}_i 2^{m+i-1}$$

Apply (7.15) in (7.14) and realize that

$$2^{2n-1} + 2^{2n-2} + 2^{2n-1} + 2^{2n-2} = 2^{2n-1} \qquad (7.16)$$

in a $2n$-bit 2's complement addition, the product becomes

$$2^{2n-1} + (\overline{a}_{n-1} + \overline{b}_{n-1} + a_{n-1}b_{n-1})2^{2n-2}$$
$$+ \sum_{i=0}^{n-2}\sum_{j=0}^{n-2} a_i b_j 2^{i+j} + (a_{n-1} + b_{n-1})2^{n-1} \qquad (7.17)$$
$$+ a_{n-1}\sum_{j=0}^{n-2}\overline{b}_j 2^{n+j-1} + b_{n-1}\sum_{i=0}^{n-2}\overline{a}_i 2^{n+i-1}$$

The term 2^{2n-1} can be created by adding a 1 to the most significant bit position of the product. The Baugh-Wooley array multiplier implements (7.17) with a full adder tree.

Modified Booth Multiplier

The modified Booth algorithm has been extensively used in multipliers with long operands (> 16 bits). Its principle is based on recoding the two's complement operand (i.e., the multiplier B) to reduce the number of partial products to be added. The reduction of partial products improves the performance of the multiplier.

The radix-2 modified Booth algorithm is based on partitioning the multiplier (B) into overlapping groups of 3 bits. This grouping is based on the transformation shown in (7.18)

$$B = -b_{n-1}2^{n-1} + \sum_{i=0}^{n-2} b_i 2^i$$
$$= -b_{n-1}2\cdot 2^{n-2} + b_{n-2}2^{n-2} + b_{n-3}2\cdot 2^{n-3} - b_{n-3}2^{n-3} + \dots$$
$$= (-2b_{n-1} + b_{n-2} + b_{n-3})2^{n-2} + (-2b_{n-3} + b_{n-4} + b_{n-5})2^{n-4} + \dots \qquad (7.18)$$
$$= \sum_{i=0}^{n/2-1} (-2b_{2i+1} + b_{2i} + b_{2i-1})2^{2i}$$

with $b_{-1} = 0$. Applying the grouping to a 16-bit multiplier produces eight groups: $b_{15}b_{14}b_{13}$, $b_{13}b_{12}b_{11}$, $b_{11}b_{10}b_9$, $b_9b_8b_7$, $b_7b_6b_5$, $b_5b_4b_3$, $b_3b_2b_1$, $b_1b_0b_{-1}$, where $b_{-1} = 0$.

The term $(-2b_{2i+1}+b_{2i}+b_{2i-1})$ in (7.18) represents five signed values {-2, -1, 0, 1, +2}. A partial product can then be generated for each three-bit group $b_{2i+1}b_{2i}b_{2i-1}$ according to the table in Fig. 7.16.

An example of multiplying two 8-bit numbers using the modified Booth algorithm is shown in Fig. 7.17. In order to add the partial products correctly, their signs must be extended. The sign extensions prevent a rectangular array to be formed for the partial product addition. A simple approach has been developed to address the sign extension problem.

For the sake of explanation, we use an example of multiplying two 8-bit numbers to demonstrate the handling of sign extension bits. Fig. 7.18 uses the letter "S" to mark the positions of the sign bits and their sign extension bits.

The processing of all the "S" bits is shown in Fig. 7.19. The calculation in Fig. 7.19 assumes that all the partial products are negative so all the "S" bits are 1's. Adding up all the "S" bits produces the sum of 10101011, which will be used as a correction vector. In fact, for n-bit multiplication, the adding of the "S" bits will produce an n-bit vector of 1010..1011.

The use of the correction vector is demonstrated in Fig. 7.20. No sign extension will be performed during the accumulation of the partial products. The following operations are used to conform to the assumption of all negative partial products. For a positive partial product, its sign bit (MSB) is flipped to a 1. For a negative partial product, its sign bit (MSB) is flipped to a 0. As shown, the correction vector is then added to the sum of the partial products to generate the correct final result.

$y_{i+1}y_iy_{i-1}$	Encoded Digit	Partial Product Generation
000	0	$0 \times A$
001	+1	$+1 \times A$
010	+1	$+1 \times A$
011	+2	$+2 \times A$
100	-2	$-2 \times A$
101	-1	$-1 \times A$
110	-1	$-1 \times A$
111	0	$0 \times A$

Fig. 7.16 Partial product generation

```
              10110101 = -75
              01010110 = +86
                 └──┘ └──┘ └──┘
    ─────────────────────────────────
    0000000010010110    100:-2A
    11111101101010      011:+2A
    111110110101        010:+1A
    1110110101          010:+1A
    ─────────────────────────────────
    1110011011001110 = -6450
```

Fig. 7.17 Modified Booth multiplication.

```
              aaaaaaaa
              bbbbbbbb
    SSSSSSSS********
    SSSSSS**********
    SSSS************
    SS**************
    ****************
```

Fig. 7.18 Sign extension in modified Booth multiplication.

```
        11111111
        111111
        1111
        11
        10101011
```

Fig. 7.19 Correction vector production.

```
        10110101 = -75
        01010110 = +86

        110010110    100:-2A
        001101010    011:+2A
        010110101    010:+1A
        010110101    010:+1A
    0011101111001110
    10101011
    1110011011001110 = -6450
```

Fig. 7.20 Adding of correction vector.

Wallace Tree Multiplier

A Wallace tree is a full adder tree structured specially for a quick addition of the partial products. Fig. 7.21 shows the construction of a Wallace tree for a 16×16 multiplier. The modified Booth algorithm is used to generate 8 partial products, which are to be added by a Wallace tree.

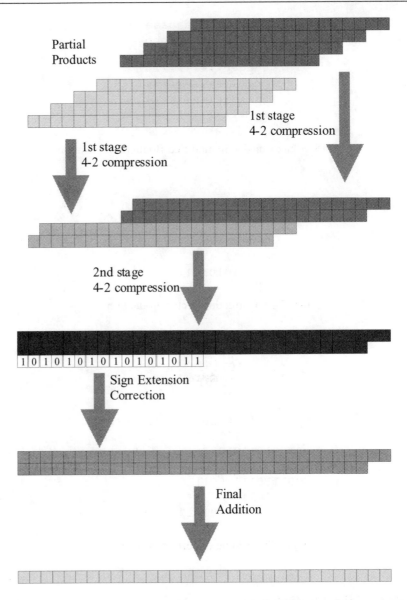

Fig. 7.21 Wallace tree multiplication.

Fig. 7.21 shows the use of 4-2 compressors to convert 4 partial products into 2 partial products. The structure of a 4-2 compressor is based on carry save addition and shown in Fig. 7.22. Each bit position of the 4-2 compressor is formed of two full-adders, which takes in 4 bits and produces 1 sum and 1 carry bits. Note that the internal carry bit does not propagate; so the delay of a 4-2 compressor is equal to that of two full adders, regardless of the length of the operands. Two parallel 4-2 compressors (stage 1) are used to convert the 8 partial products into 4 partial products. A second stage 4-2 compressor is used to further reduce the 4 partial products into 2 partial products. The sign extension correction vector is added into the partial products using the carry save addition shown in Fig. 7.23. A carry save adder compresses three inputs into two outputs. A final addition stage, usually a carry-select adder, adds up the two final partial products to produce the product.

inputs

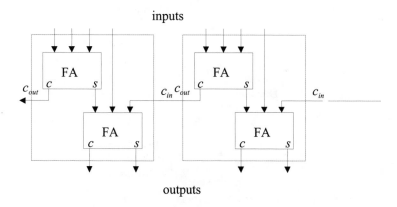

outputs

Fig. 7.22 4-2 compressor.

inputs

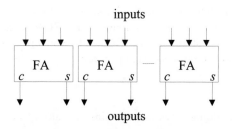

outputs

Fig. 7.23 Carry save adder.

7.4   Read-Only Memory

In contrast to the simple D-latches and D-flip-flops introduced in Chapter 5, the design of high performance memories is extremely involved. High performance memories are beyond the scope of this book. Memory module generators are commercially available for the creation of on chip memories. We present the basic structures of both read-only memory (ROM) and random access memory (RAM) to introduce their operating principles. A block diagram of a $2^n \times 1$ bit memory system is shown in Fig. 7.24. The memory array consists of 2^k rows of 2^{n-k} storage cells. A row decoder addresses one row of bits out of 2^k rows. The column decoder selects 1 bit out of the 2^{n-k} bits of the accessed row. A simple modification to the column decoder would allow m bits to be addressed together as a word.

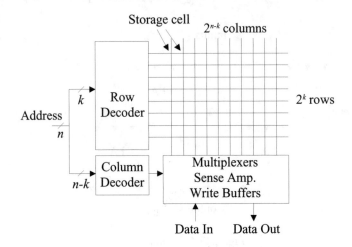

Fig. 7.24 Memory structure.

ROM can be read but not written. The stored values are created in the design process and cannot be changed. We can implement a ROM with a structure similar to a PLA. The data to be stored in the ROM are converted into sum-of-products expressions. We demonstrate this design procedure in the following example.

Example 7.6

Design a ROM building block that contains the values listed in Fig. 7.25.

The required ROM is implemented as the PLA shown in Fig. 7.26. The and-plane of the PLA in Fig. 7.26 implements a 3-to-8 decoder to select one of the rows in the or-plane which implements an 8×2 read-only memory. The decoder produces 8 product terms, which are the min-terms of the three address bits $(a_2a_1a_0)$. The or-plane simply indicates, for each output (d_1, d_0), which min-term corresponds to an output of 1. The reader should verify that the structure in Fig. 7.26 contains the same values as the data table. Instead of an and-plane structure, a complementary logic or pass-transistor logic can also be used to build the decoder.

Address $(a_2a_1a_0)$	Data (d_1d_0)
000	00
001	01
010	01
011	10
100	01
101	10
110	10
111	11

Fig. 7.25 Data values to be stored in a ROM.

7.5   Random Access Memory

While the design of high-speed, high-density RAM circuits is quite difficult, it is relatively easy to create a small RAM cell array in an ASIC. In Chapter 5 we introduced the use of D-latches and D-flip-flops to store state variables. A RAM cell, which holds one bit of memory, is based on the same principle of these circuits. However, special considerations must be applied to the implementation of storage cells in a RAM array. For example, consider the fact that a RAM cell has to be duplicated many times in the array; it is imperative that its size and power dissipation be minimized.

Fig. 7.27 shows a typical static RAM cell. The circuit is in fact a modified D-latch equipped with 2 access transistors connected to the word line (W). An address decoder (not shown) accepts an address and activates a word line to turn on the access transistors of a selected RAM cell. The access transistors, when turned on, connect the latch to the B and $\overline{B}$ lines.

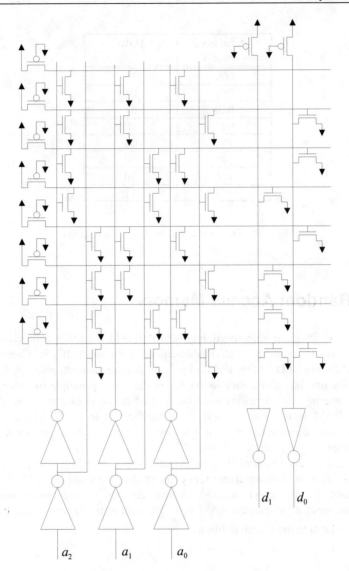

Fig. 7.26 Implementation of a ROM.

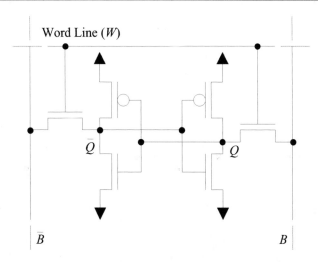

Fig. 7.27 RAM cell.

Assume that the RAM cell holds a 1 so $Q = 1$ and $\overline{Q} = 0$. The cell is in a stable state. The read operation begins by precharging lines B and $\overline{B}$ to $V_{DD}/2$ (or V_{DD}). The word (W) line is then asserted so the access transistors are turned on. Since $Q = 1$, line B is charged up through the access transistor. On the other side of the cell, $\overline{Q} = 0$ and line $\overline{B}$ is discharged. In summary, the voltage on line B will rise and that on line $\overline{B}$ will fall. Thus, a differential voltage develops between these two lines.

A differential amplifier called a sense amplifier in a RAM circuit detects this voltage difference (typically < 0.2 V) to produce a corresponding big voltage swing on lines B and $\overline{B}$. Fig. 7.28 shows the structure of a sense amplifier.

In order to minimize the size and power dissipation of a RAM cell, an individual cell does not provide sufficient charging/discharging currents to set B and $\overline{B}$ quickly into their final values. The function of the sense amplifier is to assist the read operation. The sense amplifier is a special latch provided with a pair of complementary transistors. These pair of transistors connect the cross-coupling inverters to V_{DD} and V_{SS} under the control of complementary signals (S and $\overline{S}$).

Since B and $\overline{B}$ are precharged to $V_{DD}/2$, the sense amplifier initially works at the high gain transition regions of their voltage transfer functions. When the pair of complementary transistors are turned on, depending on the voltage difference between B and $\overline{B}$, the sense amplifier latch will quickly move to one of its two stable states. Since a sense amplifier is shared by many RAM cells, we can afford to size it appropriately so that it provides adequate driving power for the read operation.

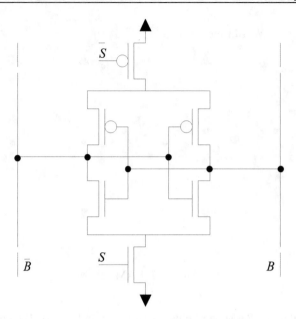

Fig. 7.28 Sense amplifier.

Now assume that a 1 is stored in the cell so $Q = 1$ and we wish to overwrite it with a 0. In the write operation, B is set to 0 and $\overline{B}$ is set to 1. The cell is selected by raising the word line. The voltage at $\overline{Q}$ is raised from 0 toward V_{DD} while the voltage at Q is being pulled down from V_{DD} toward 0. The regenerative feedback that causes the latch to switch will begin when either one reaches the logic threshold of the inverters (e.g., $V_{DD}/2$). Once this happens, the positive feedback takes over and the latch moves into its new state. Notice that the lines B and $\overline{B}$ have capacitance much higher than the inverters, so they are able to force the inverters into flipping state.

The RAM cell we have just described is a static RAM (SRAM) cell, which has the benefit of speed. However, the SRAM cell is larger in size and uses more power. Another type of RAM cell uses the same principle of a dynamic D-latch. This dynamic RAM (DRAM) cell is based on a single transistor design. The DRAM cell is thus smaller and uses less power. It is also slower and requires the stored value to be periodically refreshed.

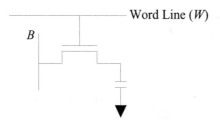

Fig. 7.29 One-transistor DRAM cell

As shown in Fig. 7.29, the single transistor DRAM cell actually contains one transistor and one capacitor. Notice that unlike a dynamic D-latch, the DRAM cell does not store its value on the gate capacitance of a transistor. The value is stored in a specially formed capacitor, which is connected to the bit line B through the transistor. The cell is selected by setting the word line W to 1. In a write operation, B is set to the new value and the capacitor is forced to the same value. In a read operation, B is first precharged before W is selected. The stored value in the capacitor causes a sense amplifier (not shown) to set the value on B accordingly.

7.6   Summary

A good VLSI design should explore modularity, regularity, and local communication. In this chapter we discussed the design of several important functional building blocks. We introduced structures to perform addition and multiplication, which are two important arithmetic functions. The implementation and operation of memory elements were also introduced. Since a function can be implemented with a number of structurally different circuits, the ability to select one that meets with desirable performance criteria is important.

7.7   To Probe Further

Algorithms

- T. H. Cormen, C. E. Leiserson, and R. L. Rivest, *Introduction to Algorithms*, McGraw-Hill/MIT Press, 1990.

Arithmetic Circuits

- K. Hwang, *Computer Arithmetic: Principles, Architecture, and Design*, John Wiley and Sons, 1979.

- J. J. F. Cavanagh, *Computer Science Series: Digital Computer Arithmetic*, McGraw-Hill, 1984.

Memory

- D. A. Hodges and H. G. Jackson, *Analysis and Design of Digital Integrated Circuits*, McGraw-Hill, 1983.

7.8 Problems

7.1   Perform an analysis on a 16-bit carry select adder to determine the optimal staging for a given technology.

7.2   Design a cell that produces the conditional sum and carry bits for the conditional sum adder.

7.3   Design an 8-bit conditional sum adder for maximum speed and minimum hardware.

7.4   Derive general equations for the number of FA's and propagation delay (in terms of FA delay time) of an m-bit, n-operand FA tree. (Hint: you may have to sacrifice some performance for regularities).

7.5   Sketch the design of a 5 bit by 5 bit Braun array multiplier.

7.6   Derive general equations for estimating the area and propagation delay of an n by m Braun array multiplier.

7.7   Prove that the operations described in (7.14) can be performed in an unsigned array multiplier provided that all partial products which involve a sign bit (a_{n-1} or b_{n-1}) and a nonsign bit are complemented. An adjustment is carried out to the final result by adding a correction term $2^{2n-1} + 2^n$.

7.8   Sketch the design of a 5 bit by 5 bit Baugh-Wooley array multiplier.

7.9   Derive general equations for estimating the area and propagation delay of an n by n Baugh-Wooley array multiplier.

7.10   Use 16×4-bit ROM's and a full adder tree to implement a multiplication function of two 4-bit fractions.

7.11   Consider the design of a processor to be used in a synchronous pipeline implementation of a digital filter. The function is to perform $Y = AX + B$.

Assume 16-bit signed operands (two's complement numbers) are used and the result is rounded up to 16 bits. Compare the relative merits of designing this processor using the following approaches:
i)   An appropriate array multiplier (Braun or Baugh-Wooley).
ii)   A ROM implementation.
iii)   A modified Booth encoded Wallace tree multiplier.
iv)   A regular add-and-shift multiplier.

7.12   Sketch a full adder tree to add $A = a_1a_0$, $B = b_2b_1b_0$, $C = c_3c_2c_1c_0$, and $D = d_4d_3d_2d_1d_0$. Minimize the delay time and the number of full adders used.

7.13   Design a circuit for producing partial products according to the modified Booth encoding algorithm.

7.14   Design a 16-bit modified Booth encoded multiplier.

7.15   Design a 16-bit modified Booth encoded Wallace tree multiplier.

Chapter 8 Chip Design

Block by block the chip is built...

This chapter illustrates the VLSI design methodology by discussing the top-down design flows of two projects: a simple microprocessor (SM) and a field programmable gate array (FPGA). Instead of a presentation of design solutions, we introduce the design flows of these projects with the expectation that the reader will assume an active role in shaping the final products. We point out critical issues in each design step and discuss potential solutions so that the reader can research the issues before moving to the next step. The quality of final products will be determined by the many decisions that a designer makes at different times during the design process.

The bottom-up flow of the design process, such as creating the layouts of functional blocks, is left for the reader to tackle. The reader can use available design tools to carry out the bottom-up design steps.

Both designs were undertaken many times as VLSI class projects. The SM project demonstrates the important concept of datapath design, in which a system is divided into a datapath unit and a control unit. We assume that the reader is reasonably familiar with the structure and operations of a microprocessor. A number of reference books are listed at the end of this chapter for those readers who need a review of this important subject.

The design of a high performance microprocessor is unarguably a very complicated task that will consume hundreds of man-months. However, our SM project is simple enough to be carried out either as a full-custom design project or as a standard-cell design project to develop essential top-down design concepts.

The second design project is to create an FPGA. We have briefly mentioned the concept of an FPGA device in Chapter 1. An FPGA consists of an array of programmable logic blocks that are connected with programmable interconnections. An FPGA can be programmed by the user in minutes with a low cost programming device. This feature has rendered FPGA a popular ASIC (application specific integrated circuit) device. CAD tools developed by the FPGA manufacturer are used to synthesize a design from its high level description into an FPGA implementation. A reference about FPGA structures is also available at the end of the chapter.

8.1   Microprocessor Design Project

Specifications:

As we have stated in Chapter 1, a project begins with the development of an abstract description for the target design. The first task is to work with the client and develop a set of specifications for the SM. At least three parameters must be decided for the SM:

- The width (i.e., number of bits) of a word.
- The size of its address space.
- The instruction set.

We arbitrarily choose a word width of 16 bits for the SM. The SM address space is chosen to be 2^{16} words, which requires the use of 16-bit addresses. Other parameters such as the SM throughput and power consumption may also be specified by the client. Throughput is typically specified in terms of MIPS (million instructions per second). It is not easy to accurately predict these properties at the initial phase of a design. Both the throughput and power consumption requirements are thus used qualitatively to guide the architecture selection in the design process. The final design will be analyzed and simulated to determine if it meets the specifications.

Points to Ponder:

- What are the implications of choosing a different word width?

- What are the implications of providing SM with a memory space of a different size?

- Suggest a few other parameters that should be included in the specifications to guide the design process.

- Define the instruction set of your SM. In order to keep the design project manageable, identify the essential instructions to be implemented. Essential instructions are operations that cannot be readily carried out by other instructions. The argument for a RISC (reduced instruction set computer) is that the microprocessor can be better optimized if it provides only a few essential instructions.

- Research and find out the definition of a low power microprocessor.

- Explain the relationship between throughput and power consumption in SM.

Architecture and Algorithm:

We partition the SM into a datapath unit and a control unit. The concept of dividing a system into a datapath unit and a control unit is essential in the design of computing structures. The datapath unit provides the engine to process (e.g., add, multiply, shift, AND, OR, etc.) the data stored in its internal registers or captured from its I/O ports. The datapath unit is a combinational circuit.

Fig. 8.1 illustrates the physical relationship between the datapath unit and the control unit, which suggests the flows of data and control signals should be routed orthogonally, probably in two different routing layers.

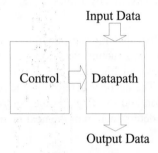

Fig. 8.1   Relationship between datapath and control units.

The setting of the datapath unit is determined by control bits supplied by the control unit in each clock cycle. Complicated operations often take multiple clock cycles to complete and require a sequence of control bits to be applied to the datapath unit. In addition, conditional operations may be involved, in which an operation depends on the result of a previous one. This suggests that the control unit is a finite state machine.

Fig. 8.2 shows the use of a finite-state machine to cycle through state sequences, each of which provides the required control bits to the datapath. The instruction stored in an instruction register sets up the control unit to produce an instruction specific sequence. The flags reflect the result of a previous operation and determine the control bit sequence for later clock cycles.

The control unit can be implemented in one of two forms. A hardwired control unit can be implemented with a sequential circuit. Programmable logic arrays (PLA) are popular for this purpose because of their capability of implementing sequential circuits with a large number of inputs, products, and outputs. The modification of a hardwired control unit (e.g., to accommodate a new instruction) is very difficult once implemented. Hard-wired control units are thus commonly used in application specific processors (e.g., digital filters) that perform fixed operations.

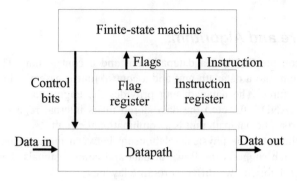

Fig. 8.2 Datapath controlled by a finite-state machine control unit.

An alternative approach called microprogramming implements the finite-state machine with a memory unit. A structure that uses a microprogrammed control unit to control a datapath is shown in Fig. 8.3. The memory stores the control bits in different control sequences. The decoder determines the next address of the memory by considering the instruction and the flags. Some control bits are also sent to the decoder to participate in the determination of the next memory address.

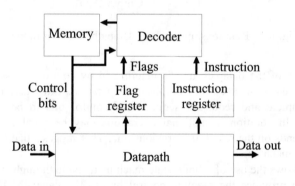

Fig. 8.3 Datapath controlled by a microprogrammed control unit.

Registers are provided in the datapath to store data fetched from memory as well as results. The information flow and processing among the data stored in the registers are called register transfer operations. A pseudo programming language called the register transfer language (RTL) is often used to describe the register transfer operations. For example, the statement $R1 \leftarrow R2 + R3$ describes that the register contents of R2 and R3 are added up and the sum is stored in register R1. Multiple register transfer operations may be carried out in the same clock period as long as they do not create a conflict on resources. Each RTL statement is an operation that can be performed in parallel on a string of bits during one clock period.

System Overview:

A top-level abstract representation of SM is shown in Fig. 8.4. The relationship between SM and a memory system is also illustrated. Many details of SM have to be specified before the top-down design flow is carried out. Fig. 8.4 shows that SM has a reset signal, which is applied from outside to put the SM into a known state.

Each of the data and address buses has 16 bits. A memory system has its own timing and handshaking requirements. These requirements are quite independent of the microprocessor. Usually the microprocessor is designed to conform to the memory system. For simplicity, we are only showing the read/write (R/W) control signal, which specifies the mode of memory access (i.e., read or write).

The reset and read/write control signals form a minimum set of microprocessor control signals. Other control signals can be added to facilitate the building of a system based on SM.

Points to Ponder:

- Study a few microprocessors and suggest other control signals that should be added to SM.

- Describe the communication between the SM and the memory system.

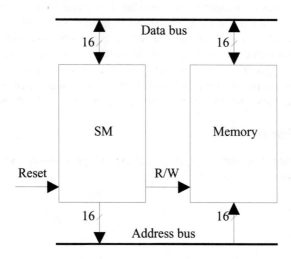

Fig. 8.4 Relationship between SM and a memory system.

The operation of SM follows that of a typical stored-program, general-purpose computer. After powered up, SM continuously monitors the reset signal. An asserted reset signal causes the microprocessor to supply a predetermined address (e.g.,

0000H[1]) on the address bus and initialize a read operation. The attached memory unit places the value stored in memory location 0000H on the data bus, which is read in, decoded, and executed. This instruction cycle is repeated until another reset signal is detected. In our example design, SM instructions consist of one to two words.

Points to Ponder:

- What are the implications of having SM instructions that spread over several words?

An example SM datapath unit is shown in Fig. 8.5. This structure represents a baseline datapath, which can use many improvements. The reader can decide whether the example datapath should be adopted as is or treated as a starting point for further developments. The structures presented in this chapter should never be considered as final designs. Standard solutions to design projects do not exist!

The core of our example datapath is an arithmetic logic unit (ALU) and a shifter, which are supported by a number of registers, a decoder, and a few multiplexers. A common clock signal synchronizes all registers.

We will describe the structure shown in Fig. 8.5 from a functional viewpoint. Two 16-bit input registers RI1 and RI2 have their inputs connected to the data bus, which is shared with the memory system (not shown). Registers RI1 and RI2 are enabled by control signals k_1 and k_2, respectively. When a register is enabled, the clock signal triggers it to capture and store the data presented on its input. General-purpose registers R1-R7 are seven 16-bit registers for storing operands and intermediate results. The inputs of all seven general-purpose registers are tied together and connected to the output of the shifter. Any data to be stored into a general-purpose register must go through the shifter, which is provided with a pass-through function for this purpose. A 3-to-8 decoder, controlled by control signal D, is used to enable at most one general-purpose register to latch in the shifter output.

Points to Ponder:

- Research and find out the typical number of general-purpose registers in a microprocessor.

---

[1] 0000H indicates a hexadecimal number 0000.

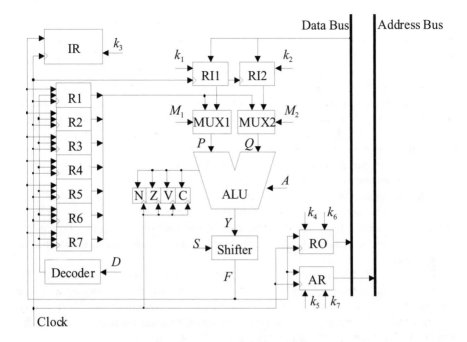

Fig. 8.5 Example SM datapath.

The multiplexer MUX1, under the control of signal M_1, chooses the input P to the ALU. Another multiplexer MUX2, under the control of signal M_2, chooses the input Q to the ALU. The selectable inputs to MUX1 are the contents of registers R1-R7 and RI1. The selectable inputs to MUX2 are the contents of registers R1-R7 and RI2.

The ALU is a multiple function combinational circuit. Control signal A determines the ALU operation performed on P and Q. An output Y is produced. Notice that there is no direct relationship between the SM instruction set and the ALU functions. Similar to the design of an instruction set, a decision has to be made on the available ALU operations. An extensive set of ALU operations simplifies the control unit design at the expense of a larger and potentially slower datapath unit. We will present as an example a small subset of possible ALU operations.

Conditional instructions (e.g., conditional branches) operate according to previous operation results. We provide four 1-bit registers (flags) to reflect the properties of result Y. The flag N indicates that Y is a negative number. The flag Z indicates that $Y = 0$. The flag V indicates that the result is too large (overflow) or too small (underflow) for Y to represent. The flag C indicates that a carry has been generated by the operation.

The shifter is controlled by signal S, which specifies the mode of shifting. We will describe a number of shifting operations, which include a pass-through (i.e., no shift) operation so that data can go through without being modified.

Register IR is an instruction register provided for the storage of an instruction brought in from the memory. Control signal k_3 enables IR to latch the output of the shifter. The value of IR is decoded by the finite-state machine control (not shown in Fig. 8.5) to produce a control sequence.

Output register RO also has its input connected to the output of the shifter. Its output is connected to the data bus. RO is used to store a result into the memory. In addition to the regular enable signal k_4 which enables register RO to capture its input, another control signal k_6 is provided to set up a tri-state connection between register RO and the data bus. This is necessary since both the datapath unit and the memory unit are sharing the data bus. When $k_6 = 0$, the output of RO is set at a high-impedance state and is isolated from the data bus.

Address register AR provides an address to the address bus for specifying the location in the memory to be accessed. The type of access, either a read or a write, is determined by setting the level of a R/W control line (not shown). Similarly, register AR has two control signals. Signal k_5 is the regular enable signal and k_7 is the tri-state control signal.

In the example SM, it controls all bus operations and its clock is also used by the memory system. A memory read cycle begins by signaling the memory system. For example, the tri-state control signal k_7 can be used for this purpose. The memory system responds by putting the data on the data bus, which is latched into SM by the next clock. The memory bus sequences are as follows:

On a read operation:
- Put an address in the AR register.
- Put the content of AR on the address bus. Signal the memory system to begin a memory read access. This can be done by combining k_7 and read/write.
- Capture the datum on the data bus into the specified register.

On a write operation:
- Put an address in the AR register.
- Put the datum to be written in the RO register.
- Put the content of AR on the address bus. Put the content of RO on the data bus. Signal the memory system to begin a memory write access. This can be done by using k_7 and read/write.

In summary, four registers, RI1, RI2, AR, and RO, provide an interface to the memory system and input/output (I/O) devices via the external address and data buses.

Fig. 8.6 describes our proposed operations of the ALU, MUX1, MUX2, shifter, and decoder. Each of the binary control signals has 3 bits and allows the selection of one out of 8 operations. This is an arbitrary choice but seems to be a reasonable selection for the complexity of a class project.

Code	A	M_1	M_2	S	D
000	$Y = P$	RI1 selected	RI2 selected	Bypass	
001	$Y = P + Q$	R1 selected	R1 selected	Logical Shift left[2]	R1 loaded
010	$Y = P + 1$	R2 selected	R2 selected	Logical Shift right	R2 loaded
011	$Y = P - Q$	R3 selected	R3 selected	Output = 0	R3 loaded
100	$Y = \overline{P}$	R4 selected	R4 selected	Arithmetic Shift Left	R4 loaded
101	$Y = P \wedge Q$	R5 selected	R5 selected	Arithmetic Shift Right	R5 loaded
110	$Y = P \vee Q$	R6 selected	R6 selected	Rotate Left	R6 loaded
111	$Y = P \oplus Q$	R7 selected	R7 selected	Rotate Right	R7 loaded

Fig. 8.6 Control signals for ALU, MUX1, MUX2, shifter, and decoder.

In Fig. 8.6, $Y = \overline{P}$ indicates that Y is the bitwise complement of P. $Y = P \wedge Q$ indicates that Y is the result of performing a bitwise-and operation between P and Q. $Y = P \vee Q$ indicates that Y is the result of performing a bitwise-or operation between P and Q. $Y = P \oplus Q$ indicates that Y is the result of performing a bitwise exclusive-or operation between P and Q.

Points to Ponder:

- Discuss the implications of using D-latches for the registers in the SM datapath unit.

- Discuss the implications of using D-flip-flops for the registers in the SM datapath unit.

- Determine the ALU operations that are adequate to implement your SM instruction set.

- Are there any other flags that should be added to the SM datapath?

---

[2] Logical shift left is identical to arithmetic shift left. In the case of shifting right, logical shift inserts a 0 as the most significant bit (MSB) of the result while arithmetic shift duplicates the MSB of the original datum as the new MSB (i.e., sign extension).

- Determine the shifter operations that should be included in your SM datapath unit.

The essence of register transfer is that each operation in the datapath involves source and destination registers. The clock period is chosen according to the slowest datapath operation. Fig. 8.7 shows a few example register transfer operations that can be performed in the example datapath. All control signals given in Fig. 8.7 are represented as binary numbers. Control signals that are not needed for an operation are marked with "-" to indicate that they are don't care values. Combinational circuits (e.g., ALU, shifter, etc.) receive don't care control signals if they are not involved with the operation. It is important to make sure that registers not involved with an operation are not accidentally enabled to overwrite its contents. The logic of a hard-wired control unit can be simplified by taking advantage of don't care values.

Operation	A	M_1	M_2	S	D	$k_1k_2k_3k_4k_5k_6k_7$
R1 ← R2 − R3	011	010	011	000	001	0000000
R2 ← Shift right (R5)	000	101	-	010	010	0000000
AR ← R7	000	111	-	000	000	0000100
Memory ← AR	-	-	-	-	000	0000001

Fig. 8.7 Datapath operation examples.

The objective of the last example operation in Fig. 8.7 is to send the address stored in register AR to the memory. We have seen that this is one of the steps to fetch either an instruction or an operand from the memory. Signal k_7 turns on the tri-state buffer between the output of register AR and the address bus. Most of the control signals are don't cares.

Points to Ponder:

- Identify the critical path(s) in the SM datapath unit.

- Design a clock system for the SM datapath unit.

We take a bit-slice approach to design the ALU. A bit-slice design approach develops a one-bit ALU building block that processes one bit of the operand. Sixteen of these building blocks are cascaded into a full width ALU. This bit-slice approach allows the design to be simplified. However, similar to the situation of a ripple carry adder, there is a performance penalty on this bit-slice design approach.

We introduce two approaches to design the bit-slice ALU. The first approach is to further divide the bit-slice ALU into an arithmetic unit and a logic unit. This concept is illustrated in Fig. 8.8. The bit-slice arithmetic unit accepts input P_i, input

Q_i, and a carry bit C_{i-1} coming from a previous stage. The ALU control signal A consists of three bits. Two bits A_0 and A_1 decide the arithmetic operation to be performed. The output of the bit-slice arithmetic unit is sent to a multiplexer. A carry output bit C_i is also produced and sent to the next stage.

A bit-slice logic unit performs logic operations on input P_i and input Q_i under the control of bits A_0 and A_1. The output of the bit-slice logic unit is also sent to the multiplexer. A third control bit A_2 selects either the arithmetic output or the logic output as the output Y_i of the bit slice.

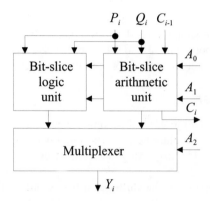

Fig. 8.8 A bit slice of the ALU.

The second bit-slice ALU structure attempts to improve the performance by designing the logic and arithmetic units as an integrated entity. If we scrutinize the logic required to perform arithmetic operations, it becomes obvious that many of the desired logic functions are implicitly included.

For example, consider the logic expressions for the sum bit (s_i) and carry output bit (c_i) of a full adder with inputs (a_i, b_i, and c_{i-1}):

$$s_i = a_i \oplus b_i \oplus c_{i-1}$$
$$c_i = a_i b_i + a_i c_{i-1} + b_i c_{i-1} \tag{8.1}$$

It is not difficult to see that if $c_{i-1} = 0$, s_i is the exclusive-or result of a_i and b_i. In addition, c_i is the logical and result of a_i and b_i. More logic functions can be extracted out of the arithmetic functions.

Points to Ponder:

- Use the 1-bit full adder described in (8.1) as the major building block to create a bit-slice ALU unit.

- Create a schematic for the bit-slice ALU shown in Fig. 8.8.

The carry-out bit of the 16-bit ALU is captured into the flag register C. The sign bit (MSB) of the result Y (2's complement number representation) is captured into the flag register N. An overflow occurs when adding two positive numbers produces a negative number or adding two negative numbers produces a positive number. A simple logic circuit that implements

$$V = C_{15} \oplus C_{14} \qquad (8.2)$$

is used to generate the signal latched into the flag register V, where C_{15} and C_{14} are the carry-outs of bit positions 15 and 14, respectively. A 16-input AND function is used to detect $Y = 0$ and its output is stored in flag register Z.

Points to Ponder:

- Consider the conditional operations (e.g., conditional branches) in a microprocessor and explain how to determine whether a condition (e.g., branch if $P > Q$) is satisfied by inspecting these four flags.

We choose to implement the shifter as a barrel shifter. Fig. 8.9 shows a partial implementation of a barrel shifter. We illustrate the use of a group of switches to implement a shifting function. The inputs ($Y_{15} - Y_0$) runs horizontally and forms a mesh with the vertical outputs ($F_{15} - F_0$). Two groups of switches are shown. One group of switches is represented by the single lines. Another group of switches is represented by the double lines. Each switch connects an input line with an output line.
If the group of single line switches is turned on, the shifter passes the input Y to output F unchanged (i.e., corresponding to $S = 000$). If the group of double line switches is turned on, the shifter rotates input Y to the right one bit position (i.e., corresponding to $S = 110$). Other groups of switches are added similarly to implement the other shifter functions. A decoder can be used to generate the switch control signals according to S. It is important that only one group of switches should be turned on at a time.

Points to Ponder

- Create an efficient layout for the shifter. Use t-gates to implement the switches.

Now we turn our attention to the design of the control. Its design depends on the choice of an instruction format. Fig. 8.10 shows an example instruction format.

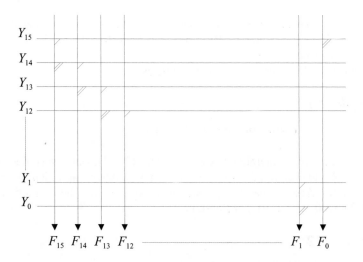

Fig. 8.9 Partial implementation of the shift register.

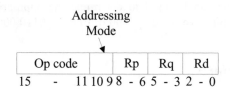

Fig. 8.10 Instruction format.

According to Fig. 8.10, the 16-bit instruction is divided into 5 fields. The opcode field (bit 15-bit 11) specifies the operation to be performed by an instruction. It has 5 bits so we can have up to 32 unique instructions. The next two bits are used to specify an addressing mode. Fig. 8.11 shows 4 example addressing modes to be used in SM. In addressing modes $(01)_2$ and $(11)_2$, the offset m or the constant will be stored in the word following the instruction.

Bits 10, 9	Addressing Mode	Address Calculation
00	(Rx)	Address = Rx
01	m(Rx)	Address = m + Rx
10	(Rx)[Ri]	Address = Rx + Ri
11	constant	Address = PC + 1

Fig. 8.11 Addressing modes.

R7 is designated as the program counter (PC). The register number Rx is stored in field Rp. The register number Ri is stored in field Rq. The field Rd specifies the destination register.

As an example, let us look at a few examples.

Example 8.1

Determine the instruction for LOAD (R2)[R3], R5.

The instruction is interpreted as R5 ← Mem[R2+R3], which specifies the values stored in R2 and R3 are added to form a memory address, at which the content is loaded into R5. Assume that the op code for LOAD is 0000. The instruction is coded into $(0000\ 0100\ 1001\ 1101)_2$ or $(049D)_{16}$.

Example 8.2

Code LOAD #10, R6 into an SM instruction.

The meaning of this instruction is to load the constant 10 into R6 (R6 ← 10). Recall that the constant 10 is stored in the location following the instruction. The addressing mode bits can be used to determine the appropriate operations. This instruction is coded into two words: $(0000\ 0110\ 0000\ 0110\ 0000\ 0000\ 0000\ 1010)_2$ or $(0606\ 000A)_{16}$.

Points to Ponder

- How do we form an instruction to perform R6 ← R1 + R2? There is no addressing mode specified for this and other similar operations.

Fig. 8.12 shows an example structure of the control unit for SM, which presents the relationship between the instruction register IR, the datapath unit, and the elements in the control unit. It is shown that the datapath unit receives the control signals A (ALU), S (shifter), D (R1-R7), M_1 (MUX1), M_2 (MUX2), and enable signals k_1-k_7. In return, the datapath provides the control unit with the flags and their complements (N, $\bar{N}$, Z, $\bar{Z}$, V, $\bar{V}$, C, and $\bar{C}$). We will explain how these flags and their complements are used in a moment.

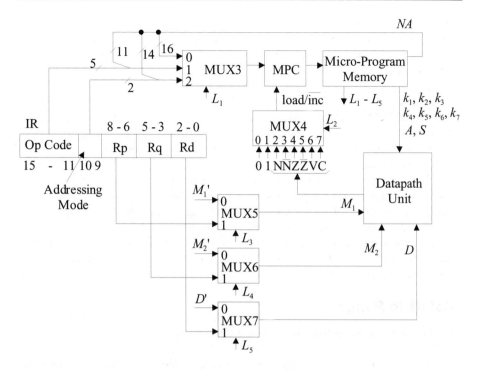

Fig. 8.12 Control unit of SM.

The control signals required for register transfer operations are stored as words in the micro-program memory. Each word is a micro-instruction that specifies operations that are to be carried out in one clock cycle. Our example micro-program memory has 16-bit addresses and thus has 2^{16} unique locations. The address of the next micro-instruction in the micro-program memory is provided by a 16-bit micro-program counter (MPC). MPC has 4 different ways of acquiring a new micro-instruction address. MPC increments its current value by 1 if the control line load/$\overline{inc}$ = 0. Otherwise, MPC loads a new address through multiplexer MUX3.

Multiplexer MUX3 has three different inputs. The first input comes entirely from the next address (NA) field in the micro-instruction being accessed. This allows the address of the next micro-instruction to be completely specified in the current micro-instruction. The second input combines 11 bits from NA and the 5-bit op code (bits 15-11) in the instruction register IR. This allows the next micro-instruction to be chosen by an instruction in IR. The third input combines 14 bits from NA and two bits (bits 10 and 9) in IR that specifies an addressing mode. This allows micro-instructions to operate differently according to their addressing modes.

Multiplexer MUX4 presents one of its 8 inputs as the MPC load/$\overline{inc}$ signal according to its control signal L_2. The output of MUX4 thus decides whether MPC should be incremented by one or loaded from MUX3. The inputs to MUX4 are provided with 2 deterministic inputs (0 and 1) so the micro-instruction being accessed can fully control the generation of the next address. Alternatively, the flags (N, Z, V, C) can be used to determine the next address so that conditional operations are possible.

Three other multiplexers MUX5, MUX6, and MUX7 provide the control signals M_1, M_2, and D to the datapath unit. MUX5, controlled by control signal L_3, chooses between IR field Rp and micro-instruction field M_1'. Similarly, MUX6 selects, under the control of L_6, either IR field Rq or micro-instruction M_2' as M_2. Multiplexer MUX7 sends, under the control of L_7, either IR field Rd or micro-instruction field D' to the datapath unit as control signal D.

The micro-program memory contains the control sequences as a micro-program. Each word of the MPC is partitioned into many fields. The use of these control signals is demonstrated by the micro-programming example shown in Fig. 8.13. In Fig. 8.13, all control signals (except k_1-k_7) and the MPC address are given in hexadecimal numbers. Control signals k_1-k_7 are given in binary numbers. Notice that additional control signals (e.g., R/W) for interfacing the memory have been omitted to simplify the discussion.

Points to Ponder:

- Develop the microprogram for your instruction set.

The microprogram memory can be implemented by means of a read-only memory (ROM). Once the design is completed, it should be simulated to verify its functions.

8.2 Field Programmable Gate Array

While a full-custom IC has the potential of achieving the best performance possible offered by a given fabrication process, its high design complexity often makes it cost prohibitive. Programmable logic devices (PLD), such as the programmable logic array (PLA) that we have introduced in Chapter 6, were developed as user-programmable, general-purpose devices to implement logic as two-level sum-of-products.

Two-level sum-of-products are inadequate for large designs. A mask-programmed gate-array, or simply gate-array, is a chip that consists of an array of prefabricated transistors. Gate-arrays are customized by wiring up the transistors to form multi-level logic, which is done at the foundry with custom metal layer masks.

	MPC Address	L_1	L_2	L_3	L_4	L_5	M_1'	M_2'	D'	NA	A	S	k_1-k_7
Instruction fetch:													
AR ← R7	0000	-	0	0	-	0	7	-	0	-	0	0	0000100
BUS ← AR; R7 ← R7 + 1	0001	-	0	0	-	0	7	-	7	-	2	0	0000001
RI1 ← BUS	0002	-	0	-	-	0	-	-	0	-	-	-	1000000
IR ← RI1	0003	-	0	0	-	0	0	-	0	-	0	0	0010000
Interpretation:													
determine next address	0004	1	1	-	-	0	-	-	0	0100	-	-	0000000
	....												
LOAD:													
addressing modes	0100	2	1	-	-	0	-	-	0	0200	-	-	0000000
(other instructions)...	....												
Mode 00:	0200	0	1	-	-	0	-	-	0	0204	-	-	0000000
Mode 01:	0201	0	1	-	-	0	-	-	0	0208	-	-	0000000
Mode 10:	0202	0	1	-	-	0	-	-	0	020C	-	-	0000000
Mode 11:	0203	0	1	-	-	0	-	-	0	020D	-	-	0000000
(Rx)													
AR ← Rx	0204	-	0	1	-	0	-	-	0	-	0	0	0000100
BUS ← AR	0205	-	0	-	-	0	-	-	0	-	-	-	0000001
RI1 ← BUS	0206	-	0	-	-	0	-	-	0	-	-	-	1000000
Rd ← RI1	0207	0	1	0	-	1	0	-	-	0000	0	0	0000000
m(Rx)													
AR ← R7	0208	-	0	0	-	0	7	-	0	-	0	0	0000100
BUS ← AR; R7 ← R7 + 1	0209	-	0	0	-	0	7	-	7	-	2	0	0000001
RI2 ← BUS	020A	-	0	-	-	0	-	-	0	-	-	-	0100000
AR ← RI2 + Rx	020B	0	1	1	0	0	-	0	0	0205	1	0	0000100
(Rx)[Ri]													
AR ← Ri + Rx	020C	0	1	1	1	0	-	-	0	0205	1	0	0000100
constant													
AR ← R7	020D	-	0	0	-	0	7	-	0	-	0	0	0000100
BUS ← AR; R7 ← R7 + 1	020E	0	1	0	-	0	7	-	7	0206	2	0	0000001
...	....												

Fig. 8.13 Microprogramming example.

A field programmable gate array (FPGA) is a general-purpose, multi-level programmable logic device. Unlike the mask-programmed gate arrays, FPGAs are programmable by their users with inexpensive software and hardware, often provided by the manufacturer of the FPGAs. A class of FPGAs, which are called SRAM-programmed FPGAs, are reprogrammable and thus reusable.[3] This feature proves to be very powerful for the debugging of a design and has made FPGAs very popular. SRAM FPGAs are programmed by loading configuration data into configuration memory. Another class of FPGAs is anti-fuse-programmed. Anti-fuse works in a manner opposite to fuses. An anti-fuse device irreversibly changes from a high to a low resistance when a programming voltage is applied.

Points to Ponder:

- From the viewpoint of area, power, and speed, compare the SRAM-based FPGA and the anti-fuse-programmed FPGA.

- An alternative to the FPGA is a device called CPLD (configurable programmable logic device). The CPLD contains an array of PLA-like structures interconnected by programmable routing resources. Research this device and discuss its advantages and disadvantages.

An SRAM-based FPGA must be reprogrammed every time it is turned on since the configuration data are lost when the power is turned off. A side effect of its volatility is that an SRAM-based FPGA system needs an external memory for storage of its configuration so that it can be initialized during power-up. The other side of being volatile is that SRAM-based FPGAs are ideal for prototyping since they can be reprogrammed without cost. The configuration memory can be replaced with EPROM or EEPROM to preserve the configuration after the power is turned off.[4]

Traditionally, the entire FPGA must be programmed all together due to the fact that the configuration data are sent in serially as a bit stream. This does not permit a part of the array to be reconfigured without affecting the rest of the circuit. Partially reconfigurable FPGAs are now also available. Since a portion of an FPGA can be selectively modified, an implementation can be customized according to the task on hand for better performance. In addition, an FPGA can be efficiently time-shared between several applications. Similar to the concept of a cache memory, an application specific configuration can be loaded into the FPGA on an as-needed basis.

The cost of a FPGA is low since it can be mass-produced. Each part can be fully tested by the manufacturers so that the yield, from the user's view point, is virtually 100%. The biggest advantage of SRAM-based FPGAs is probably in the fact that they are always fabricated with the most advanced technology developed for SRAM. All logic of an FPGA is implemented in static logic. Provided with a means to turn off the unused portion of the FPGA, the power dissipation can be kept reasonably low and the standby current is virtually zero.

---

[3] SRAM: static random access memory.
[4] EPROM: Electrically programmable read-only memory. EEPROM: Electrically erasable and programmable read-only memory.

Our second design example is an SRAM-based FPGA. Fig. 8.14 shows a general floorplan of an SRAM-based FPGA. It contains an array of configurable logic blocks (CLBs) connected with configurable interconnects (routing resources). These programmable routing resources are provided to connect the CLBs together to form the desired function. A ring of programmable I/O blocks are arranged along the edges of the FPGA.

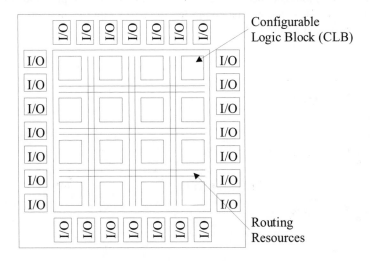

Fig. 8.14 General floorplan of an FPGA.

A CLB allows an arbitrary logic function of its inputs to be implemented. A single bit register can be incorporated into a CLB so that sequential circuits can be readily constructed. The configurable interconnects allow the CLBs to be connected to implement large, multi-level logic. The programmable I/O blocks can be programmed to be an input port, an output port, or a bi-directional port, or a tri-state port. They provide the communications between the pins of an FPGA chip and the internal circuit. Special CAD tools have been developed to synthesize FPGA designs from high-level behavioral description.

Points to Ponder:

• Design the specifications of a programmable I/O block.

• How does the rest of the circuit affect the design of the I/O block?

• Should a latch or a flip-flop be used to implement the single bit register?
The building blocks of an FPGA CLB include a lookup table (LUT), programmable interconnect points (PIPs), a 1-bit register, several multiplexers, and a configuration memory. An example structure of a CLB is shown in Fig. 8.15 along

with the routing resources. Note that the configuration memory is not shown in Fig. 8.15.

The LUT can be used to implement any logic function of four inputs by storing its truth table. Four multiplexers selectively connect the inputs of LUT to the vertical routing interconnects so that the LUT can receive outputs of other CLBs. The output of the LUT is sent to a single-bit register, which is represented as a D-flip-flop. A 2-to-1 multiplexer chooses either the LUT output or the register output as the output of the CLB. The CLB output can be connected to one or more of the vertical routing interconnects by setting the PIPs.

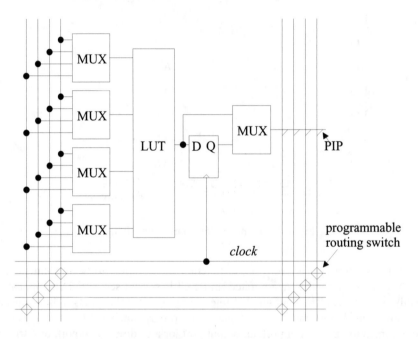

Fig. 8.15 Block diagram of CLB and its relationship with the routing resource.

The four-input LUT can be implemented by a 16×1 SRAM memory. The memory is used as a function generator to implement any of 2^{16} logic functions of 4 inputs. The truth table of the desired function is loaded into the memory when the CLB is configured. Since there is no difference between the four inputs to the LUT, the assignment of input signals to physical input pins can be done in a way to optimize the routing of signals.

The programmable routing switches provide the programmable connections between the horizontal and vertical routing interconnects. The structure of a programmable routing switch is shown in Fig. 8.16. The six PIPs are used to configure a number of routing modes. In addition to providing the connection between a horizontal routing path to a vertical routing path, the programmable routing switch also allows a routing interconnect to be broken into two sections. This way different sections of the same routing interconnect can be used for different routing paths. Each PIP is controlled by the output of a one-bit memory cell. If a 1 is stored in the control memory cell, the PIP turns on and completes a connection.

Since each routing interconnect has to go through a number of PIPs, the delays introduced by them can become significant. The long wire also contributes large loading capacitance to the CLB.

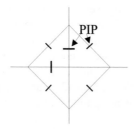

Fig. 8.16 Programmable routing switch.

Another chip level circuitry is the circuit for loading the configuration data to the LUT and the control memory. The configuration data can be stored in registers. A simple implementation is to connect all configuration registers into a shift register so that the configuration data are loaded serially as a long control word. Alternatively, the configuration memory can be implemented as a random access memory so that a portion of the configuration data can be loaded without overwriting the rest. This capability is required by a partially reconfigurable FPGA.

Points to Ponder:

- The structure in Fig. 8.15 shows that each CLB has four inputs. Research commercial FPGA products and determine the number of inputs for your CLB. What are the implications of having more or less inputs?

- Instead of running long routing interconnects all across the chip as shown in Fig. 8.15, some FPGA designs have provided routing wires of different lengths. Shorter wires are used for local interconnections and longer wires are used for global interconnections. Design a similar scheme for your FPGA routing resources.

- Discuss the pros and cons of using pass-transistors to implement the PIP.
- What are the implications of using a random access memory to implement the configuration memory?

- The propagation delay due to an interconnect is significant because of the use of PIPs to provide reconfigurability. Long wires should be buffered. The designer may decide to provide buffering in the interconnect. Instead of providing dedicated buffers for the long routing interconnects, a signal can be routed through a CLB to achieve a similar effect. Discuss the advantages and disadvantages of this approach.

In addition to the common criteria (i.e., area, speed, and power) that we use to evaluate a design, a general-purpose logic device should also be evaluated by considering its logic density and routability. The logic density is defined as the amount of usable logic per unit chip area.

The configurable routing resource does not contribute to the logic density. Theoretically the availability of more metal layers allows the wiring to be routed on top of the configurable logic blocks instead of around them. However, the fact that PIPs are needed to provide the programmable connections limits this option since devices cannot overlap.

If the goal is to increase the logic density, minimizing routing resources may seem to be an attractive option. However, a high raw density does not imply a high usable density. The lack of routing resources renders many logic blocks inaccessible.

A higher degree of routability can be provided by adding more PIPs. However, each PIP requires the area of a pass transistor plus that of a memory cell. These overheads add to the size of the chip and reduce the density. As mentioned before, these pips also have adverse effects on the signal speed.

Points to Ponder:

- The highest clock frequency that an FPGA can operate is determined by the delay of a CLB and the propagation delay of sending the signal to a neighbor CLB. Use this observation to guide the design of your FPGA.

8.3   Summary

In this chapter, we go through the top-down flow of two designs. The first design is a simple microprocessor that demonstrates the concept of the datapath and control approach. The second design is an FPGA, which represents a class of popular user programmable logic devices.

8.4   To Probe Further

Microprocessor Design:

- J. L. Hennessy and D. A. Patterson, *Computer Organization and Design: The Hardware/Software Interface*, Morgan Kaufmann Publishers, 1994.

- M. M. Morris, *Computer System Architectures*, 3rd edition, Prentice Hall, 1993.

Field-Programmable Logic Arrays:

- S. M. Trimberger, ed., *Field-Programmable Gate Array Technology*, Kluwer Academic Publishers, 1994.

- *IEEE Design and Test of Computers*, January – March, 1998. This is a special edition on FPGAs.

Low Power Design:

- A. Bellaouar and M. Elmasry, *Low-Power Digital VLSI Design Circuits and Systems*, Kluwer Academic Publishers, 1995.

- G. K. Yeap, *Practical Low Power Digital VLSI Design*, Kluwer Academic Publishers, 1998.

Commercial FPGAs:

- http://www.xilinx.com

- http://www.altera.com

- http://www.actel.com

8.5   Problems

8.1 Complete the design of the SM.

8.2 Complete the design of the FPGA.

8.3 Design an alarm clock.

8.4 Design a floating point multiplier.

8.5 Design a jogging meter that calculates and records the jogging distance.

8.6 Design a bicycle meter that calculates traveling speed.

8.7 Design a BCD (0000 - 1001) counter with a 7 segment display output.

Chapter 9 Testing

Final judgment ...

The testing of a chip is an operation in which the chip-under-test is exercised with carefully selected test patterns (stimuli). The responses of the chip to these test patterns are captured and analyzed to determine if it works correctly. A faulty chip is one that does not behave correctly. The incorrect operation of a chip may be caused by design errors, fabrication errors, and physical failures, which are referred to as faults. Fig. 9.1 lists a few example faults in each of these three categories.

Design errors	Incomplete specifications
	Incorrect logic implementations
	Incorrect wiring
	Design rule violations
	Excessive delays
	Glitches or hazards
	Slow rise/fall times
	Improper noise margins
	Improper timing margins
Fabrication errors	Shorts
	Opens
	Improper doping profiles
	Mask misalignments
	Incorrect transistor threshold voltages
Physical failures	Electron migration
	Cosmic radiation and α-particles

Fig. 9.1 Example faults found in integrated circuits.

In some cases, we are only interested in whether the chip-under-test behaves correctly. For example, chips that have been fully debugged and put in production normally require only a pass-or-fail test. The chips that fail the test are simply discarded. We refer to this type of testing as fault detection.

In order to certify a prototype chip for production, the test must be more extensive in nature to exercise the circuit as much as possible. The test of a prototype also requires a more thorough test procedure called fault location. If incorrect behaviors are detected, the causes of the errors must be identified and corrected.

An important problem in testing is test generation, which is the selection of test patterns. A common assumption in test generation is that the chip-under-test is non-redundant. A circuit is non-redundant if there is at least one test pattern that can distinguish a faulty chip from a fault-free one.

A non-redundant combinational circuit with n inputs is fault-free if and only if it responds to all 2^n input patterns correctly. Testing a chip by exercising it with all its possible input patterns is called an exhaustive test. This test scheme has an exponential time complexity so it is impractical except for very small circuits.

For example, 4.3×10^9 test patterns are needed to exhaustively test a 32-input combinational circuit. Assume that we have a piece of automatic test equipment (ATE) that can feed the circuit with test patterns and analyze its response at the rate of 10^9 patterns per second (1 GHz).[1] The test will take only 4.3 seconds to complete, which is long but may be acceptable. However, the time required for an exhaustive test quickly grows as the number of inputs increases. A 64-input combinational circuit needs 1.8×10^{19} test patterns to be exhaustively tested. The same piece of test equipment would need 570 years to go over all these test patterns.

The testing of sequential circuits is even more difficult than combinational circuits. Since the response of a sequential circuit is determined by its operating history, a sequence of test patterns rather than a single test pattern would be required to detect the presence of a fault. There are also other problems in the testing of a sequential circuit, such as the problem of bringing the circuit into a known state and the problem of timing verification.

The first challenge in testing is thus to determine the smallest set of test patterns that allows a chip to be fully tested. For chips that behave incorrectly, the second challenge is to diagnose, or locate, the cause of the bad response. This operation is difficult because many faults in a chip are equivalent; so they are indistinguishable by output inspection. For example, if the output of an inverter incorrectly produces a 0, the following faults are indistinguishable without a further investigation:

- There is a short between the inverter output and V_{SS}.
- There is a short between the inverter input and V_{DD}.
- The pull-down transistor (nMOS) of the inverter is permanently turned on.
- ...

Fortunately, a truly exhaustive test is rarely needed. In addition, it is often sufficient to determine that a functional block, instead of an individual signal line or transistor, is the cause of an error. We begin with the discussion of popular fault models that allow practical test procedures to be developed.

---

[1] This would be one of the top-of-the-line testers in 2000. The cost of such a tester is around 2 million US Dollars.

9.1   Fault Models

As noted above, except for very small circuits, it is impractical to pursue an exhaustive test. Instead, a test should consist of a set of test patterns that can be applied in a reasonable amount of time. This test should provide the user with the confidence that the chip-under-test is very likely to be fault free if it passes the test. An important issue in the development of a test procedure is thus to evaluate the effectiveness of a test. The quality of a test can be judged by an index called fault coverage. Fault coverage is defined as the ratio between the number of faults a test detects and the total number of possible faults. This is usually determined by means of a simulated test experiment. This experiment, which is called a fault simulation, uses a software model of the chip to determine its response to the test when faults are present. A fault is detected by a test pattern if the circuit response is different from the expected fault-free response.

Fault models are created to facilitate the generation of test patterns. A fault model represents a subset of the faults that may occur in the chip-under-test. Several fault models have been developed for representing faults in CMOS circuits. These models can be divided into logic fault models, delay fault models, and current-based fault models.

The most widely used logic fault models are the stuck-at fault, stuck-open fault, and bridging fault models. Delay fault models incorporate the concept of timing into fault models. Examples of delay fault models are the transition delay and path delay fault models. The current-based fault models were developed by recognizing the very low leakage current of a CMOS circuit. Many defects, such as opens, shorts, and bridging, result in a significantly larger current flow in the circuit.

Many automatic test pattern generation (ATPG) algorithms are based on the "single fault assumption." This assumption assumes that at most one fault exists at any time so the test generation complexity can be significantly reduced. We will revisit this subject later in this chapter. Brief introductions to these fault models are given below. We will discuss some of these models in detail.

Stuck-at Fault Model

The stuck-at fault model assumes that a design error or a fabrication defect will cause a signal line to act as if it were shorted to V_{SS} or V_{DD}. If a line is shorted to V_{SS}, it is a constant 0 and is named a stuck-at-0 (s-a-0) fault. On the other hand, if a line is shorted to V_{DD}, it is a constant 1 and is called a stuck-at-1 (s-a-1) fault. The stuck-at fault model is most effective if it is used at the inputs and outputs of a logic unit such as a logic gate, a full adder, etc. The application of this fault model in a test is to force a signal line to 1 for a s-a-0 fault and to 0 for a s-a-1 fault. The response of the circuit is then analyzed.

Stuck-Open Fault Model

The stuck-open (s-op) fault model attempts to model the behaviors of a circuit with transistors that are permanently turned off. The result of having transistors that would not be turned on is unique to CMOS circuits. A s-op fault changes a CMOS combinational circuit into a sequential circuit. A two-step test is thus required. The first step brings the signal line being tested to an initial value. The second step then carries out the test in a way similar to the testing of stuck-at faults.

Bridging Fault Model

A bridging fault model represents the accidental connection of two or more signal lines in a circuit. The most common consequence of a bridging fault is that the shorted signal lines form wired logic so the original logic function is changed. It is also possible that the circuit becomes unstable if there is an unwanted feedback in the circuit. Bridging faults can be tested by applying opposite values to the signal lines being tested.

Transition Delay Fault Model

The transition delay fault model is based on the stuck-at model. The application of this fault model is similar to the stuck-at model except that the s-a-0 and s-a-1 values are replaced by 0-to-1 and 1-to-0 transitions. An observed signal line is forced to a known value at some time before it is observed. A transition is then applied and the signal line is observed along with a time assessment. Possible faults are slow-to-rise (0-to-1 transition) faults and slow-to-fall (1-to-0 transition) faults. This fault model is commonly used to model a functional block that has an excessive delay.

Path Delay Fault Model

The path delay fault model considers the total transition delay of a circuit path. The only difference between the path delay fault model and the transition delay model is that an entire path instead of a single signal node is targeted. The same operation of assigning an initial value and observing a transition is used.

Current-Based Fault Model

Some faults cannot be detected by either the logical fault model or the delay fault model. These defects, which do not cause a logical error but represent a reliability problem, may cause a high leakage current. This defect induced leakage current may be an order of magnitude larger than the normal CMOS quiescent current. The current-based fault model is only applicable when the chip is in a static state.

Our discussion begins with the stuck-at and stuck-open fault models. Fig. 9.2 shows a two-input NAND gate that is used to explain the modeling of physical defects as stuck-at and stuck-open faults. We label the four transistors in this circuit as Q_{pA}, Q_{pB}, Q_{nA}, and Q_{nB}, respectively.

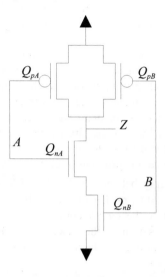

Fig. 9.2 CMOS NAND gate.

The stuck-at fault model is often used along with the further assumption that there is at most one fault in the circuit, which is known as the single-fault assumption. As mentioned above, the stuck-at fault model is often applied to the inputs and outputs of a logic block. For a logic block with n inputs and m outputs, $2(n + m)$ single stuck-at faults are possible. If we consider only the inputs and output of this NAND gate, six single stuck-at faults are possible:

- A: s-a-0.
- A: s-a-1.
- B: s-a-0.
- B: s-a-1.
- Z: s-a-0.
- Z: s-a-1.

By definition, if input A: s-a-1, the NAND gate sees $A = 1$, regardless of the value placed on it. Many stuck-at faults of a circuit are equivalent and thus indistinguishable. For example, input A: s-a-0 is equivalent to and thus indistinguishable from output Z: s-a-1. When equivalent stuck-at faults are taken into consideration, the number of single stuck-at faults in the NAND gate is reduced to four: A: s-a-1, B: s-a-1, Z: s-a-1, and Z: s-a-0.[2]

A stuck-open fault occurs if transistor Q_{nA} is permanently opened. A possible cause of this stuck-open fault is that either its source or drain is disconnected from

---

[2] Z: s-a-1 is equivalent to either A: s-a-0 or B: s-a-0.

the rest of the circuit. In the presence of Q_{nA}: s-op, when $AB = 11$, output Z becomes a high impedance state (HiZ) since it is connected neither to V_{SS} nor to V_{DD}. Output Z thus retains its previous logic value; so the circuit acts like a sequential circuit. The length of time before the value at Z changes is determined by the capacitance of node Z and the leakage current.

The stuck-at and stuck-open fault models do not cover all physical faults in a CMOS circuit. For example, consider a bridging between the source and drain of transistor Q_{nA}. This is equivalent to a situation in which transistor Q_{nA} is turned on all the time. When $AB = 01$, a voltage divider is formed by transistors Q_{pA} and Q_{nB}, which drives output Z to an intermediate value between V_{DD} and V_{SS}. The value perceived at Z is determined by the logic threshold of the stage accepting it as input.

Bridging faults are often difficult to detect. We use a slightly more complicated logic function to demonstrate this problem. Fig. 9.3 shows a CMOS implementation of $Z = \overline{AB + CD}$. A bridging between nodes 1 and 2 causes the Z output to be intermediate when $ABCD = 1001$ or $ABCD = 0110$. If the stage driven by Z perceives this intermediate value as 1, the circuit continues to function as expected with a substantially smaller noise margin.[3] Otherwise, if Z is below the logic threshold of the next stage, the bridging fault has converted the function to $Z = \overline{A(B + D) + D(B + C)}$.

Fig. 9.4 compares the NAND gate (Fig. 9.2) outputs at a few example faulty conditions with their corresponding fault-free outputs. The last column (Q_{nA}: bridged) depicts the output when the source and drain of transistor Q_{nA} are shorted. In the next section, we will show that test patterns can be generated by comparing the faulty and fault-free outputs of a circuit.

9.2   Test Generation (Stuck-at Faults)

Test generation deals with the selection of input combinations that can be used to verify the correct operation of a chip. Consider a circuit with a function $F(X) = F(x_1, \ldots, x_n)$, where X is an input vector representing n inputs $x_1, \ldots, x_n$.

Suppose we would like to find a test pattern to detect a single stuck-at fault occurring at an internal circuit node k (i.e., $k \notin X$). The first observation is that node k must be set to 0 in order to detect k: s-a-1 or 1 to detect k: s-a-0. The second observation is that a test pattern X_k qualified to detect the specific k stuck-at fault must satisfy the following condition. When X_k is applied to the circuit, the fault-free response $F(X_k)$ must be different from the incorrect output $F'(X_k)$ caused by the stuck-at fault at k. This is the basic principle of test generation.

---

[3] This is the kind of fault in which the current-based fault models are most useful. A larger than normal current is produced by the short.

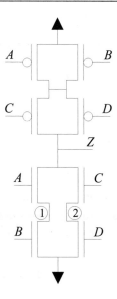

Fig. 9.3 CMOS implementation of $Z = \overline{AB + CD}$.

AB	Z (fault-free)	Z (A: s-a-1)	Z (A: s-a-0)	Z (Q_{nA}: s-op)	Z (Q_{nA}: bridged)
00	1	1	1	1	1
01	1	0	1	1	X
10	1	1	1	1	1
11	0	0	1	HiZ	0

Fig. 9.4 A few possible faults in a CMOS NAND gate.

Normally it is impossible to directly inject a value at an internal node of a chip. It is thus necessary to find an input combination X_k that can set k to the desired value. If we can set the value of a node of a chip, either directly in the case of an input node, or indirectly in the case of an internal node, the node is said to be controllable.

Unlike a board-based design, it is impractical to physically probe the internal nodes of a chip for their values. In order to observe an internal node, some path must be chosen to propagate the effect of a fault to the chip output. The test pattern X_k must be chosen to sensitize a path from the node under test to an observable output. If the value of a node can be determined, either directly in the case of an output, or indirectly in the case of an internal node, it is said to be observable.

Now we formalize the requirement of a test pattern that detects a stuck-at fault at an input x_i. X_i is a test vector for detecting x_i: s-a-1 if and only if

$$\overline{x}_i \bullet (F(X_i) \oplus F'(X_i)) = 1 \tag{9.1}$$

and a test vector for detecting x_i: s-a-0 if and only if

$$x_i \bullet (F(X_i) \oplus F'(X_i)) = 1 \tag{9.2}$$

where $F(X_i) = F(x_1,...,x_i,...,x_n)$ and $F'(X_i) = F(x_1,...,\overline{x}_i,...,x_n)$.

In (9.1), the term $\overline{x}_i$ ensures that x_i is set to 0. Similarly, in (9.2), the term x_i ensures that x_i is set to 1. The exclusive-or term used in (9.1) and (9.2) is called the Boolean difference of $F(X)$ with respect to its input x_i and can be written as

$$\frac{dF(X)}{dx_i} = F(x_1,...,x_i,...,x_n) \oplus F(x_1,...,\overline{x}_i,...x_n) \tag{9.3}$$

which specifies that the variables other than x_i must be assigned values so that the output is sensitive to a change of x_i.

The principles specified in (9.1) and (9.2) can be generalized to specify test patterns for an internal node of a combinational circuit. This can be easily done by rewriting $F(x_1,...,x_n)$ as $F(x_1,...,x_n,k)$, in which k is the internal node for which a test pattern is to be determined. The test pattern requirements are then generalized as follows.

X_k is a test pattern for detecting k: s-a-1 if and only if

$$\overline{k}\frac{dF(X)}{dk} = 1 \tag{9.4}$$

and a test pattern for detecting k: s-a-0 if and only if

$$k\frac{dF(X)}{dk} = 1 \tag{9.5}$$

As an example, consider a logic function $F = x_1 x_2 + x_3 x_4$. Assume that $k = x_1 x_2$ is an internal node of the circuit. We can rewrite the function as $F = k + x_3 x_4$ and $k = x_1 x_2$. The tests for k: s-a-1 are found by considering

$$\bar{k}\frac{dF}{dk} = \overline{x_1 x_2} ((k + x_3 x_4) \oplus (\bar{k} + x_3 x_4))$$
$$= \overline{x_1 x_2} (\overline{x_3 x_4})$$
(9.6)
$$= \overline{x_1}\overline{x_3} + \overline{x_2}\overline{x_3} + \overline{x_1}\overline{x_4} + \overline{x_2}\overline{x_4}$$
$$= 1$$

The following test patterns $x_1 x_2 x_3 x_4 = $ 0-0-, -00-, 0--0, -0-0, in which the '-' indicates a don't care value, satisfy (9.6) and are thus the tests for k: s-a-1.

The tests for k: s-a-0 are found by considering

$$k\frac{dF}{dk} = x_1 x_2 ((k + x_3 x_4) \oplus (\bar{k} + x_3 x_4))$$
$$= x_1 x_2 (\overline{x_3 x_4})$$
(9.7)
$$= x_1 x_2 \overline{x_3} + x_1 x_2 \overline{x_4}$$
$$= 1$$

An analysis of (9.7) yields test patterns $x_1 x_2 x_3 x_4 = $ 110- and 11-0 for k: s-a-0.

The above test generation principles have been implemented in various approaches. All these approaches are based on the assumption that the circuit-under-test is non-redundant and has at most a single stuck-at fault. The single-fault assumption may be justifiable for a fully debugged chip coming out of a production line. This assumption does not apply to a prototype chip which may have more than one fault caused by design errors or fabrication defects. However, most automatic test pattern generation algorithms still adopt the single fault assumption since the determination of test patterns can be significantly simplified. In practice, many multiple faults will also be detected by a test set generated under the single fault assumption. With the exception of stuck-open faults, a test set generated by considering single stuck-at faults may also cover other faults such as the bridging faults. Faults that are not detected in a fault simulation can be considered individually so that their test patterns can be generated to enhance the test set.

9.3   Path Sensitization

Test generation involves two steps: fault activation and error propagation. Fault activation requires setting the circuit primary inputs so that a s-a-v line has a value $\bar{v}$. Error propagation seeks primary input values to propagate the resulting error to a primary output. Path sensitization is a direct implementation of (9.4) and (9.5). If the fault locates at an internal node of the circuit, a difference at the node being tested must be created. For example, a test vector that attempts to detect k: s-a-0 must set k to 1. A sensitized path must be found to propagate the difference from its origin to the output. The necessary conditions to create the difference at the tested node and to propagate the fault along the sensitized path are then established.

Path sensitization can be applied as a manual approach to identify test vectors for small circuits. The next section explains a computer-aided test generation algorithm that implements the concept of path sensitization.

9.4   D-Algorithm

D-algorithm is the pioneer of many computer-aided test generation methods. D-algorithm uses symbols D and $\bar{D}$ to represent errors. If we use D to denote a 0/1 error (0 is the expected value and 1 is the observed value), then $\bar{D}$ denotes a 1/0 error (1 is the expected value and 0 is the observed value). The meanings of D and $\bar{D}$ can be exchanged as long as their uses are consistent throughout a chip-under-test. Error-free values 0/0 and 1/1 are simply denoted by 0 and 1, respectively. Adding an unspecified (don't care) value X, D-algorithm performs test generation by carrying out 5-valued logic operations in the chip-under-test. The 5-valued logic operations are shown in Fig. 9.5 (see Problem 9.2).

AND	0	1	D	$\bar{D}$	X
0	0	0	0	0	0
1	0	1	D	$\bar{D}$	X
D	0	D	D	0	X
$\bar{D}$	0	$\bar{D}$	0	$\bar{D}$	X
X	0	X	X	X	X

OR	0	1	D	$\bar{D}$	X
0	0	1	D	$\bar{D}$	X
1	1	1	1	1	1
D	D	1	D	1	X
$\bar{D}$	$\bar{D}$	1	1	$\bar{D}$	X
X	X	1	X	X	X

Fig. 9.5   5-valued logic operations in D-algorithm.

Consider the problem of generating a test of c: s-a-0 in the 2-input NAND gate shown in Fig. 9.6. The behavior of this faulty NAND gate is represented by the truth table of Fig. 9.7, in which the X's indicate don't care values. This truth table simply says that output c remains 0 regardless of the values of a and b.

Fig. 9.6   2-input NAND gate.

In order to detect c: s-a-0, we need to set c at 1 to create a D (or $\bar{D}$, as long as it is consistent throughout the circuit). The input pattern (ab) can be easily determined by selecting one from the NAND gate's fault-free truth table that produces $c = 1$. Three patterns ($ab = 00$, 01, and 10) are possible.

a	b	c: s-a-0
X	X	0

Fig. 9.7 Truth table of a NAND gate with its output s-a-0.

Compact truth tables called singular covers are used in the D-algorithm. The truth table of a logic gate can be simplified by incorporating the don't care value (X). A singular cover of a logic gate can be generated by inspecting any two rows in the original truth table with identical outputs. In this inspection, any input on which the output does not depend is marked as a don't-care (X). The results of these inspections are collected to form the gate's singular cover. Fig. 9.8 shows the singular cover for a two-input NAND gate. According to the singular cover of a two-input NAND gate, the input patterns that set $c = 1$ are $0X$ and $X0$.

a	b	$\overline{ab}$
0	X	1
X	0	1
1	1	0

Fig. 9.8 Singular cover for two-input NAND gate.

A pattern formed by an input combination of a logic circuit and the logic circuit's response to this input combination is called a cube. For example, the rows ($0X1$, $X01$, 110, 001, etc.) in the singular covers shown in Fig. 9.8 are cubes. A primitive D-cube of a fault is a cube that brings the effect of a fault to the output of the logic circuit. It is used to generate a difference (i.e., D) at the faulty node to be tested.

In the ongoing example of determining a test pattern for c: s-a-0 (Fig. 9.6), if we set the inputs of the NAND gate to $ab = 0X$ or $X0$, $c = D$. The primitive D-cubes for c: s-a-0 are thus $0XD$ and $X0D$.

A primitive D-cube for a logic function can be constructed by selecting one cube from the fault-free singular cover and one cube from the singular cover of the faulty circuit, which should have different output values. These two cubes are then intersected according to the intersecting rules given in Fig. 9.9, which describes the result of intersecting two values in corresponding positions of two cubes.

Intersect ($\wedge$)	0	1	X
0	0	$\bar{D}$	0
1	D	1	1
X	0	1	X

Fig. 9.9 Intersecting rules.

Applying this intersection operation to $XX0$ (a cube from the faulty NAND gate with c: s-a-0, see Fig. 9.7) and $0X1$ (a cube from the fault-free NAND gate, see Fig. 9.8), we have primitive D-cube $0X\bar{D}$ (or $0XD$). Similarly, intersecting $XX0$ and $X01$ produces primitive D-cube $X0\bar{D}$ (or $X0D$). This result is consistent with the one found by observation.

A primitive D-cube can also be found for a faulty input of a logic function. Suppose we would like to find a primitive D-cube for b: s-a-0 for the 2-input NOR gate shown in Fig. 9.10. The singular cover of the NOR gate with b: s-a-0 is shown in Fig. 9.11. The singular cover of the fault-free NOR gate is shown in Fig. 9.12.

Fig. 9.10 Two-input NOR gate.

a	b	c
0	X	1
1	X	0

Fig. 9.11 Singular cover for NOR gate with b: s-a-0.

a	b	c
0	0	1
1	X	0
X	1	0

Fig. 9.12 Singular cover for fault-free NOR gate.

The primitive D-cube for b: s-a-0 is generated as follows. Since b: s-a-0, it must be set to 1 to create a difference at b. Cube $X10$ fits with this description and is selected. It is then intersected with cube $0X1$ from the faulty gate singular cover. This produces primitive D-cube $01D$.

D-algorithm uses propagation D-cubes to sensitize a path which propagates the difference D or $\bar{D}$ caused by a fault to a primary output. Propagation D-cubes can be found by inspecting a gate's singular cover. All cubes that cause the output to depend only on one or more of its inputs are propagation D-cubes.

The propagation D-cubes of a logic function can be systematically constructed by intersecting cubes with different output values in its singular cover. For example, the propagation D-cubes of a 2-input NAND gate (Fig. 9.6) is $abc = 1D\bar{D}$, $D1\bar{D}$, and $DD\bar{D}$.

The use of D-algorithm to determine test patterns follows the steps shown below.

1) Select a primitive D-cube for the fault of which test vectors are to be determined.

2) Select propagation D-cubes from the logic gates in the path from the faulty node to the output. This allows the difference (D or $\bar{D}$) to be propagated to the output so that it can be observed. This is called the forward trace operation.

3) For all other logic blocks that are not involved with the sensitized path, try to match the cubes in their singular cover with the values determined so far. A consistent set of input values is the valid test vector. If a consistent set of input values cannot be found, no test vector can be found for this fault (e.g., the circuit is redundant).

An example is used here to demonstrate the use of D-algorithm to identify test vectors.

Example 9.1

Use D-algorithm to generate test patterns for g: s-a-1 in the circuit shown in Fig. 9.13.

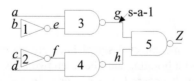

Fig. 9.13 Example circuit for D-algorithm.

The signal line g is the output of a 2-input NAND gate (gate 3). The primitive D-cube for gate 3 is thus selected to be $aeg = 11D$. The D at g must be propagated to the primary output Z through gate 5. Gate 5 has propagation D-cubes $ghZ = 1D\bar{D}$, $D1\bar{D}$, etc. We select $D1\bar{D}$ as the propagation D-cube of gate 5 to match with the primitive D-cube of gate 3. The rest of the signals are selected from the singular covers of gates 1, 2, 4 to be consistent with the signals determined so far. The steps of D-algorithm are shown in Fig. 9.14. Notice that the selection of a cube in each step must be consistent with the values selected in previous steps. The test patterns are found to be $101X$ (i.e., 1010 and 1011). Other test patterns can be found by selecting a different singular cover cube for gate 4.

	a	b	c	d	e	f	g	h	Z
Primitive D-cube (gate 3)	1				1		D		
Propagation D-cube (gate 5)							D	1	$\bar{D}$
Singular cover (gate 1)		0			1				
Singular cover (gate 4)				X		0		1	
Singular cover (gate 2)			1			0			

Fig. 9.14 D-algorithm.

9.5   Test Generation for Other Fault Models

Stuck-Open Faults

Recall that a stuck-open fault transforms a CMOS combinational circuit into a sequential circuit. In order to detect a stuck-open fault, the observable node must be first driven to a known initial value. Consider finding the test sequence for Q_{nA}: s-op in the NAND gate in Fig. 9.2. Setting $AB = 00$, 01 or 10 will drive output Z to an initial value of 1 when a second test vector of $AB = 11$ is applied. A fault free circuit produces $Z = 0$ in response to this test sequence. On the other hand, $Z = 1$ when Q_{nA}: s-op.

Bridging Faults

When two normally unconnected signal lines are shorted, we have a bridging fault. A general model for a bridging fault between two lines a and b is shown in Fig. 9.15. Once a bridging fault occurs between signals a and b, these values become unobservable. We consider the values of a and b in the model their driven values, not observed values.

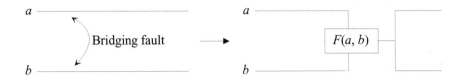

Fig. 9.15 Bridging fault model.

If *a* and *b* are identical, the function $F(a, b)$ assumes the same value. When *a* and *b* have opposite values, the value $F(a, b)$ is indeterminate.[4] This situation can be verified by considering two inverters with their outputs tied together (see problem). Indeterminate signal values are very difficult to detect since its value may depend on the following stage's logic threshold.

If there exists at least one path between the bridged lines, the short causes a feedback bridging fault. A combination circuit can be converted into a sequential circuit by the presence of a feedback bridging fault and thus requires a test sequence to detect the fault.

However, it is easy to show that a bridged signal driven by opposite signals causes an abnormal current to flow through the circuit, which can be detected by a current-based test.

9.6 Test Generation Example

The task is to generate a set of test patterns for the full adder shown in Fig. 9.16.

Fig. 9.16 Full adder.

The full adder circuit has 3 inputs (*a*, *b*, and *c*) and 2 outputs (*d* and *e*). Outputs *d* and *e* are its carry and sum outputs, respectively. Since it has 3 inputs, it can be exhaustively tested by all 8 possible input combinations from 000 to 111. Assume that the single stuck-at fault model is used to determine a set of test patterns.

---

[4] This statement is only valid in a CMOS circuit. In other technologies (e.g., TTL), a short between two lines form a hardwired AND or OR.

The fault list of the full adder is:
- a: s-a-1.
- b: s-a-1.
- c: s-a-1.
- d: s-a-1.
- e: s-a-1.
- a: s-a-0.
- b: s-a-0.
- c: s-a-0.
- d: s-a-0.
- e: s-a-0.

Fig. 9.17 lists all test patterns for each of these stuck-at faults. The fault coverage of each test pattern is summarized in the fault matrix shown in Fig. 9.18. Inspecting the fault matrix reveals that we only need two patterns $abc = 000$ and 111 to detect any single stuck-at faults at the inputs and outputs of the full adder.

It is a big reduction of test vectors obtained according to the single stuck-at fault model. But, how well does this test do when it is used in practice? For the sake of this example, we assume that the full adder is implemented by the circuit shown in Fig. 9.19, which is a typical standard cell implementation of the full adder.

Signal m is not directly accessible. We will determine a test vector to detect m: s-a-0. We need a test vector that will set m to 1, an opposite value of its stuck-at fault. Input vector 000 will do that. Since signal m is not directly observable, it must be propagated to either output d or e. In order to reflect any change of signal m at output d, the remaining 2 inputs of NAND gate must be set to 1, which is also achieved by test vector 000. In other words, vector 000 detects this internal fault m: s-a-0.

In order to detect a fault m: s-a-1, m must be set to 0. Vector 111 satisfies this requirement. However, it does not provide the necessary values on the other inputs of the NAND gate to propagate the change at m to output d. So the fault m: s-a-1 is not detectable by the test vectors determined by considering single stuck-at faults at the inputs and outputs of the full adder. The detection of m: s-a-1 requires the inputs of the NAND gate that produces d to be 011.

The above example demonstrates the attempt to identify a minimum number of tests to verify the correctness of a chip. First, a number of critical nodes (e.g., the inputs and outputs of the full adder in Example 9.2) are selected to generate test vectors. This produces a set of test vectors. A process called fault simulation is then performed to evaluate the fault coverage of this test set. Faults not considered by the initial fault model (e.g., the internal stuck-at faults and stuck-open faults in Example 9.2) are injected into the circuit simulated by a circuit simulator. We call this a fault simulation. The test vectors are applied to the simulated faulty circuit to determine if the fault introduced can be detected by at least one of them. If it does, the fault is covered. Otherwise, new test vectors can be added to enhance the fault coverage, which is defined to indicate the percentage of faults that are detected by the test vectors.

Fault	Test Patterns (*abc*)	Fault-free output (*de*)	Faulty output (*de*)
a: s-a-1	000	00	01
	001	01	10
	010	01	10
	011	10	11
b: s-a-1	000	00	01
	001	01	10
	100	01	10
	101	10	11
c: s-a-1	000	00	01
	010	01	10
	100	01	10
	110	10	11
a: s-a-0	111	11	10
	100	01	00
	101	10	01
	110	10	01
b: s-a-0	111	11	10
	010	01	00
	011	10	01
	110	10	01
c: s-a-0	111	11	10
	001	01	00
	011	10	01
	101	10	01
d: s-a-1	000	00	10
	001	01	11
	010	01	11
	100	01	11
e: s-a-1	110	10	11
	000	00	01
d: s-a-0	011	10	00
	101	10	00
	110	10	00
	111	11	01
e: s-a-0	001	01	00
	010	01	00
	100	01	00
	111	11	10

Fig. 9.17 Test patterns for all stuck-at faults in a full adder.

Test vector	a s-a-1	b s-a-1	c s-a-1	a s-a-0	b s-a-0	c s-a-0	d s-a-1	e s-a-1	d s-a-0	e s-a-0
000	1	1	1				1	1		
001	1	1				1	1	1		
010	1		1		1		1	1		
011	1				1	1				
100		1	1	1			1	1		
101		1		1		1				1
110			1	1	1		1			1
111				1	1	1			1	1

Fig. 9.18 Fault matrix.

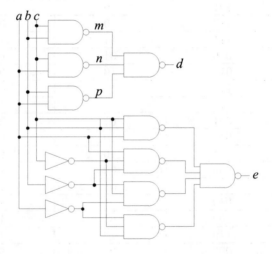

Fig. 9.19 A gate level implementation of a full adder.

9.7   Sequential Circuit Testing

Testing sequential circuits is difficult because their behaviors depend not only on present input values but also on past inputs. Conceptually, a sequential circuit can be modeled as a sequence of identical combinational circuits. Techniques developed for combinational circuit test generation can then be applied. This approach is illustrated in Fig. 9.20. We represent the sequential circuit with n identical combinational circuits. The ith combinational circuit receives input $x(i)$ and state $y(i - 1)$. The output $z(i)$ is observable. Therefore, the ith combinational circuit corresponds to the sequential circuit at the ith clock cycle.

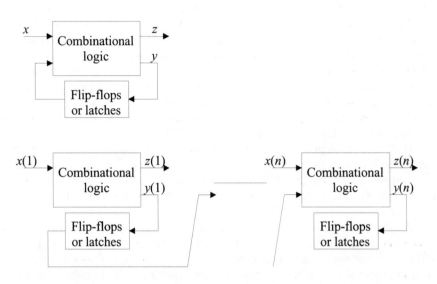

Fig. 9.20 Sequential circuit modeled as combinational circuit for test generation.

A fault occurring in the original sequential circuit transforms into n identical faults in the combinational circuit model; so it has to be treated as a multi-fault detection problem. This technique is thus only realizable for sequential circuits with a few states.

Techniques have been developed to simplify the testing of sequential circuits by increasing in testability (i.e., controllability and observability). The next section describes a number of design-for-testability approaches.

9.8   Design-for-Testability

A VLSI chip naturally has limited controllability and observability. One principle in which all IC designers agree is that a design must be made testable by providing adequate controllability and observability.[5] These properties must be well planned for in the design phase of the chip and not as an afterthought. This practice is referred to as design-for-testability (DFT).

The testability of a circuit can be improved by increasing its controllability and observability. For example, the test of a sequential circuit can be significantly

---

[5] Unfortunately, many designers choose to ignore this principle since design-for-testability adds to their design effort.

simplified if its state is controllable and observable. If we make the registers storing the state values control points, the controllability of the combinational logic's "hidden" inputs is improved. On the other hand, if we make the flip-flops observation points, the observability of the combinational logic's "hidden" outputs is increased.

This is usually done by modifying the registers so that they doubled as test points. In a test mode, the registers can be reconfigured to form a scan register (i.e., shift register). This allows test patterns to be scanned in as well as responses to be scanned out. A single long scan register may cause a long test time since it takes time to scan values in and out. In this case, multiple scan registers can be formed so that different parts of the circuits can be tested concurrently. Even though a scan-based approach is normally applied to the registers required in the function, additional registers can be added solely for the purpose of DFT.

IEEE has developed a standard (IEEE Std1149.1) for specifying how circuitry may be built into an integrated circuit to provide testability. The circuitry provides a standard interface through which communication of instructions and test data are done. This is called the IEEE Standard Test Access Port and Boundary-Scan Architecture.

Another problem of a sequential circuit testing is that we need to bring the circuit into a known state. If the initialization (i.e., reset) of a circuit fails, it is very difficult to test the circuit. Therefore, an easy and fool-proof way to initialize a sequential circuit is a necessary condition for testability. The scan-based test-point DFT approach allows registers to be initialized by scanning in a value.

If a circuit incorporates free-running clock generators or pulse generators, it is extremely hard to test. A solution is to provide a means to turn off these circuits and provide the necessary signals externally.

A number of other DFT techniques are also possible. These include the inclusion of switches to disconnect feedback paths and the partitioning of large combination circuits into small circuits. Remember the cost of testing a circuit goes up exponentially with its number of inputs. For example, partitioning a circuit with 100 inputs into 2 circuits, each of which has 50 inputs, can reduce the size of its test pattern space from 2^{100} to 2^{51} (2×2^{50}).

Most DFT techniques usually require additional hardware to be included to the design. This modification affects the performance of the chip. For example, the area, power, number of pins, and delay time are increased by the implementation of a scan-based design. A more subtle point is that DFT increases the chip area and logic complexity, which may reduce the yield. A careful balance between the amount of testability and its penalty on performance must be applied.

9.9   Built-In Self-Test

Built-in self-test (BIST) is the concept that a chip can be provided with the capability to test itself. There are several ways to accomplish this objective. One way is that the chip tests itself during normal operation. In other words, there is no need to place the chip under test into a special test mode. We call this the on-line BIST.

We can further divide on-line BIST into concurrent on-line BIST and non-concurrent on-line BIST. Concurrent on-line BIST performs the test simultaneously with normal functional operation. This is usually accomplished with coding techniques (e.g., parity check). Non-concurrent BIST performs the test when the chip is idle.

Off-line BIST tests the chip when it is placed in a test mode. An on-chip pattern generator and a response analyzer can be incorporated into the chip to eliminate the need for external test equipment. We discuss a few components that are used to perform off-line BIST below.

Test patterns developed for a chip can be stored on chip for BIST purposes. However, the storage of a large set of test patterns increases the chip area significantly and is impractical. A pseudo-random test is carried out instead. In a pseudo-random test, pseudo-random numbers are applied to the circuit under test as test patterns and the responses compared to expected values. A pseudo-random sequence is a sequence of numbers that is characteristically very similar to random numbers. However, pseudo-random numbers are generated mathematically and are deterministic. This way the expected responses of the chip to these patterns can be predetermined and stored on chip. We discuss the structure of a linear feedback shift register shortly, which can be used to generate a sequence of pseudo-random numbers.

The storage of the chip's correct responses to pseudo-random numbers also has to be avoided for the same reason of avoiding the storage of test patterns. An approach called signature analysis was developed for this purpose. A component called a signature register can be used to compress all responses into a single vector (signature) so that the comparison can be done easily. Signature registers are also based on linear feedback shift registers.

Linear Feedback Shift Register

Linear feedback shift registers (LFSR) are used in BIST both as a generator of pseudo-random patterns and as a compressor of responses. Consider the feedback shift register shown in Fig. 9.21, which illustrates the sequence it generates. Each box represents a flip-flop. The flip-flops are synchronized by a common clock and form a rotating shift register. Assume that the initial value in the shift register is 110. It is shown that the shift register goes through a 3-pattern sequence $110 \rightarrow 011 \rightarrow 101$. The sequence repeats afterward.

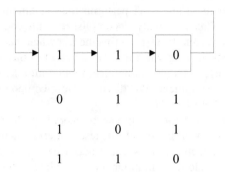

Fig. 9.21 Shift register with feedback.

A shift register with n flip-flops can produce at most 2^n different patterns. This is easy to see if we connect the n flip-flops into an n-bit counter.

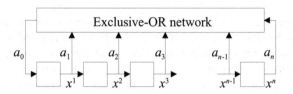

Fig. 9.22 Linear feedback shift register.

Fig. 9.22 shows the general structure of a linear feedback shift register. Each box represents a D-flip-flop. All flip-flops are synchronized by a common clock (not shown). The exclusive-or network in the feedback path performs modulo-2 addition of the values (x's) in the flip-flops. The value at each stage is a function of the initial state of the linear feedback shift register and of the feedbacks. A linear feedback shift register with n stages can be described by its characteristic polynomial:

$$P(x) = a_n x^n + a_{n-1} x^{n-1} + \ldots + a_2 x^2 + a_1 x^1 + a_0 \tag{9.8}$$

in which a_i ($i = 1$ to n) is either 1 or 0 and $a_0 = 1$.

By suitably choosing the characteristic polynomial, we can build an n-stage linear feedback shift register, which when initialized to a nonzero value runs autonomously and generates an output sequence whose period is 2^n-1. A linear feedback shift register that implements $P(x) = x^3 + x + 1$ has three stages with the following settings:

$a_3 = 1$.
$a_2 = 0$.

$a_1 = 1$.
$a_0 = 1$.

This linear feedback shift register produces a sequence of 7 non-zero binary patterns as shown in Fig. 9.23.

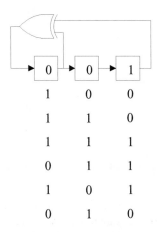

1	0	0
1	1	0
1	1	1
0	1	1
1	0	1
0	1	0

Fig. 9.23 Linear feedback shift register as a pseudo-random pattern generator.

Signature Analyzer

The idea of a signature analyzer is to compress a sequence of data into a unique code. It uses a linear feedback shift register provided with an extra input in the exclusive-or network to accept a sequence of bits to be compressed. After the entire sequence has been clocked through, the state of the shift register is called a signature. Apparently the signature depends on the initial state of the shift register as well as the input sequence. In most cases the shift register is initialized to be all zeroes.

We modify the linear feedback shift register shown in Fig. 9.23 to provide an external input to the exclusive-or feedback network. The result, which is a three-bit signature analyzer, is shown in Fig. 9.24. It is shown that two sequences, supposedly coming from a circuit under test, produce different signatures after they are clocked through the linear feedback shift register. It can be seen that the signature can be used to indicate the presence of a fault. Instead of storing and comparing a long sequence of data, a signature is all that is needed to carry out a built-in self-test.

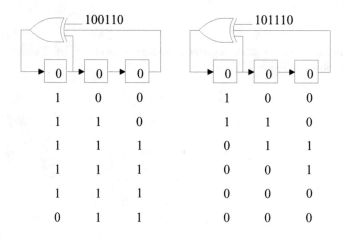

Fig. 9.24 Signature analyzer.

When a sequence of n bits is encoded by an m-bit signature ($m < n$), more than one sequence will map into one signature. There are 2^n unique sequences and 2^m unique signatures in this situation. In average, each signature will represent $2^n/2^m = 2^{n-m}$ sequences. The probability of declaring an incorrect sequence correct since it produces the expected signature is

$$\frac{2^{n-m} - 1}{2^n - 1} \qquad (9.9)$$

The denominator in (9.9) is the number of incorrect sequences. The numerator is the number of incorrect sequences that would map into the signature identical with that of the correct sequence. Normally, $n \gg m > 1$; so (9.9) can be approximated as

$$\frac{2^{n-m}}{2^n} = 2^{-m} \qquad (9.10)$$

The probability of drawing an incorrect conclusion from using a signature analyzer can then be made arbitrarily small by choosing a large m. Normally $m = 16$ would give an acceptable error probability. When the signature is incorrect, the circuit is not functioning properly. If the signature is correct, we can only conclude that the circuit has a high probability to be functioning correctly.

Multiple data sequences can be combined and compressed with a signature analyzer with multiple inputs to produce a multiple-input signature. We conclude this section with Fig. 9.25, which shows the use of a pseudo-random pattern generator and a signature analyzer to test a circuit.

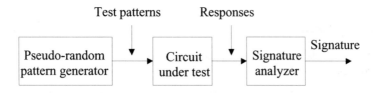

Fig. 9.25 Setup of a pseudo-random pattern generator and a signature analyzer to test a circuit.

9.10 Summary

The complexity of exhaustively testing a chip grows exponentially with the number of inputs. A number of fault models were introduced for the generation of test patterns. Among them, the most popular fault model is the single stuck-at fault model. This chapter described the determination of test patterns for stuck-at faults. The concepts of fault simulation and fault coverage were introduced. The testability of a chip can be increased by improving its controllability and observability. The use of linear feedback shift registers to build pseudo-random pattern generators and signature analyzers for built-in self test was explained.

9.11 To Probe Further

D-Algorithm:

- J. P. Roth, "Diagnosis of automata failures: a calculus and a method," *IBM Journal of Research and Development*, vol. 10, no. 7, July 1966, pp. 278-291.

Boolean Difference:

- F. F. Sellers, M. Y. Hsiao, and C. L. Bearnson, "Analyzing errors with the Boolean difference," *IEEE Trans. on Computer*, July 1968, pp. 676-683.

Pseudo-Random pattern generator:

- P. H. Bardel, W. H. McAnney, and J. Savir, *Built-in Test for VLSI: Pseudorandom Techniques*, NY: John Wiley & Sons, Inc., 1987.

Boundary Scan:

- http://standards.ieee.org/reading/ieee/std_public/description/testtech/1149.1-1990_desc.html

- *IEEE Standard Tests Access Port and Boundary-Scan Architecture,* IEEE Standard 1149.1 1990, IEEE Standards Board, 1990.

- K. P. Parker, *The Boundary-Scan Handbook, 2nd Edition, Analog and Digital,* Kluwer Academic Publishers, 1998.

Design-For-Testability:

- L. Crouch, *Design for Test For Digital IC's and Embedded Core Systems,* Prentice-Hall, 1999.

- M. Abromovici, M. A. Breuer, and A. D. Friedman, *Digital Systems Testing and Testable Design,* Computer Science Press, 1990.

Built-In Self-Test:

- J. Rajski and J. Tyszer, *Arithmetic Built-In Self-Test for Embedded Systems,* Prentice-Hall, 1998.

IDDQ Test:

- R. K. Gulati and C. F. Hawkins, eds., IDDQ Testing of VLSI Circuits–A Special Issue of Journal of Electronic Testing: Theory and Applications, Kluwer Academic Publishers, 1995.

9.12 Problems

9.1   Find the pseudo-random sequences in 4-bit LFSRs defined by the following polynomials. (a) $x4+x3+x2+1$; (b) $x4+x2+x$; (c) $x4+x3+1$; (d) $x4+x3+x2$.

9.2   Verify the 5-value logic operations for D-algorithm given in Fig. 9.5.

9.3   Develop a test set that detects all single stuck-at faults in Fig. 9.26.

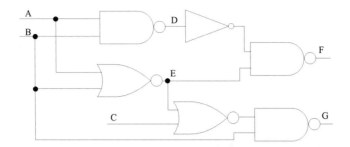

Fig. 9.26 Circuit for Problem 9.3.

9.4   Find the singular cover for a logic function $Z = \overline{a \cdot b + c}$.

9.5   Find the propagation D-cube for the logic function $Z = \overline{a \cdot b + c}$.

9.6   Find the primitive D-cube for $Z = \overline{a \cdot b + c}$ when Z: s-a-1.

9.7   Show that $\dfrac{d(F \oplus G)}{dx} = \dfrac{dF}{dx} \oplus \dfrac{dG}{dx}$.

Chapter 10 Physical Design Automation

Means to an end ...

Sophisticated computer-aided design (CAD) tools have become the necessities of IC designers to deal with the complexity of today's VLSI systems. Physical design process is a collective term indicating the operations involved in the production of layout (mask) information for the fabrication of VLSI systems. We go behind the scene in this chapter to explore the CAD tools used in the physical design process.

A non-exhaustive list of physical design tools includes layout editors, layout generators, design rule checkers (DRC), layout-versus-schematics verifiers (LVS), circuit extraction tools, and place-and-route tools. Layout editors and generators aid the time consuming and error prone layout production process. They are used to create library cells, and, in some situations, complete full-custom chip layouts. Design rule checkers and layout-versus-schematics verifiers ensure layout correctness. Circuit extraction tools generate simulation models from layouts. Place-and-route tools convert building block netlists (e.g., high-level synthesizer results) into physical layouts.

10.1 Layout Generators and Editors

In an ASIC design environment, structural representations are generated from behavioral models. The output of a synthesis system is typically a netlist of standard-cells. Place-and-route tools are then used to assemble these standard-cells into a layout. We explain place-and-route tools in Section 10.2.

As explained in Chapter 3, standard-cells are limited to relatively simple logic functions. The performance of standard-cells are inherently limited since they are designed to be used in a range of applications. In order to improve the performance of a standard-cell based ASIC, the same logic function can be designed in different ways to accommodate different applications. For example, different versions of a standard-cell logic function can be designed to provide different driving capabilities.

Certain higher level functions (e.g., random access memory, multiplication, etc.) have structural regularity. These regularly structured library cells can be generated on-the-fly when they are needed. Layout generators, also called module generators, use templates and leaf cells (basic cells) to generate regular parameterized layouts.

Consider an array multiplier as an example (see Chapter 8). It is less practical to store multipliers of all different sizes such as 8-bit by 8-bit, 16-bit by 16-bit, etc. Instead, an array multiplier template, which specifies the number of full adders (leaf cells) and their interconnections, can be used with a layout generator to create multiplier layouts of desired sizes.

Layout generators are specialized tools. Targeting toward the parameterized production of a single circuit structure allows a layout generator to provide optimized results. In contrast, layout editors are general-purpose tools used to design library cells and leaf cells. A layout editor provides its user with an interactive color-painting environment to create geometrical shapes. Different colors are used to represent different mask layers. Some systems allow only Manhattan geometry, which is inspired by the North-South, East-West street layout in Manhattan, New York, whereas others allow the use of arbitrary angles (non-Manhattan geometry). A rectilinear grid is usually super-imposed on the layout area for the easy drawing of precise shapes. Shapes can be selected, grouped, copied, and pasted. They can also be rotated, mirrored, and scaled.

Basic place-and-route functions could be provided in a layout editor to assemble cells into hierarchical layouts. Other useful layout editor features include design rule checking (DRC), electrical rule checking (ERC), layout compaction, and circuit extraction. The public domain layout editor *magic* (see Chapter 3) provides incremental DRC capability to interactively check for design rule violations while shapes are being added to a layout. In the design of a large circuit, performing DRC interactively may take away computing power from other activities such as the display of a large layout. Most design methodologies thus treat DRC as a post-layout process. Tools have also been developed to perform ERC to detect electrical errors such as the use of polysilicon (instead of metal) as power rails and inadequate substrate/well contacts. In sub-micron designs, layouts should also be checked to verify that they do not violate minimum density rules and antenna rules (see Section 3.2).

Circuit extraction is an important layout function for design verification. A circuit extraction process determines the transistor dimensions, parasitic capacitance/resistance, and connectivity of a layout under design. The extraction of parasitic capacitance/resistance is a difficult problem, especially in deep sub-micron designs (i.e., 0.25 μm or below). In deep sub-micron designs, the capacitance between interconnects becomes dominant (see Section 4.5).

Some circuit extraction tools flatten the entire circuit (i.e., remove the hierarchy) before it is extracted. More advanced circuit extraction tools perform the extraction hierarchically. An extracted circuit that preserves its original hierarchy provides a more efficient model for further operations (e.g., simulation). In addition to providing a model for simulation, an extracted netlist can be verified against a schematic by an LVS tool.

Layout compaction attempts to move the geometrical objects in a layout to minimize the layout area, while enforcing design rule correctness. The use of layout compaction is not limited to reducing the overall chip area by removing unnecessary spaces in a layout. Since design rules must be enforced in the compaction process, compaction has also been applied as the basic technology in the migration of a layout to a different technology. For example, a design created for fabrication process *A* can be migrated into a layout that conforms with the design rules of fabrication process *B*. While there are still limitations in the capability of compaction tools, their uses have become increasingly popular as the industry emphasizes the importance of reuse. For example, compaction tools have been used to migrate standard-cell libraries to a new fabrication technology.

The movement of an object in a compaction process is naturally a two-dimensional problem. Two-dimensional movements provide a compaction process the potential of achieving better results; however, it is a very hard problem. Most layout compaction algorithms thus break the problem into two one-dimensional compaction problems. First, the compaction is performed along one direction (e.g., horizontal), followed by a compaction in the perpendicular direction. All interconnections are assumed stretchable. Smallest movable objects are identified and shifted by the correct amount without breaking any connections. User specified constraints such as transistor dimensions are preserved.

10.2 Placement and Routing

Modern VLSI designs rely heavily on library cells. The layout process is thus simplified to the arrangement and wiring of required library cells. Placement arranges a set of building blocks (e.g., standard cells) on a chip surface. Routing defines wiring paths to interconnect electrically equivalent building block pins.

Placement and routing, both combinational in nature, are hard problems. An optimal solution can only be reached by exhaustively searching the solution space of the problem. An exhaustive search problem cannot be solved in polynomial time with a single processor. Computer scientists call this type of problems "NP-complete" problems or "NP-hard" problems.[1] Many placement and routing tools are thus based on "practical" heuristic algorithms to deal with the very large number of building blocks in VLSI circuits.

These heuristic algorithms have been the subjects of many research efforts. Insights to their operations and limitations are valuable to the use of these heuristic and thus non-deterministic tools.

10.3 Floorplanning and Placement

According to the divide-and-conquer design technique, a large VLSI circuit is somehow partitioned into subcircuits. The physical design process thus views the circuit under design as a set of building blocks interconnected by a set of signal nets. The objective of placement is to optimally arrange this set of building blocks within a limited chip area subject to various positional and electrical constraints.

The arrangement of building blocks is usually done in two phases. A preparation phase, called floorplanning, is often applied to define the building block and thus the layout hierarchy. This can be viewed as a generalized placement, which is commonly used to determine the shape and dimensions of each building block by rearranging its components. The overall layout area can be estimated in this phase.

The placement phase specifies exact building block locations. The placement result is fed to a routing tool to complete the interconnections. Floorplanning and

---

[1] NP stands for non-polynomial.

placement are similar in many aspects. The following discussion refers to both as "placement" in the most general sense. When adequate, differences between floorplanning and placement are pointed out.

The input of a placement process is usually a set of *n* rectangular building blocks. A feasible placement solution divides a placement area into *n* non-overlapping slots for placing the building blocks. Fig. 10.1 shows several feasible placements for a design case of four building blocks.

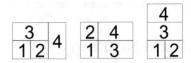

Fig. 10.1 Several feasible placements for four building blocks.

The number of feasible placements for a given problem is very large, even for a small number of building blocks. Consider the simplified placement problem illustrated in Fig. 10.2. Six building blocks (1–6) are to be placed into an area divided into 6 equally sized slots (A–F). Each slot can hold any one building block. The total number of feasible placements for this simple problem is $6! = 6 \times 5 \times 4 \times 3 \times 2 \times 1 = 720$.

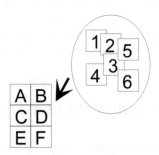

Fig. 10.2 Simple placement problem.

In order to select good placements from many feasible solutions, an objective function can be used to measure the quality of placements. First, a placement must allow the required routing to be achieved within a given area. Since modern design methodology often treats placement and routing as two separate processes, routing space must be reserved by the placement algorithm. It is not uncommon that a seemingly good placement failing the 100% routable requirement is ripped up and redone. A placement may be measured by its overall area and overall interconnection length. The delays of critical circuit paths can also be used as a criterion. An objective function may also include penalty terms to represent violations such as overlapping building blocks.

Before routing, overall area and overall interconnection length are not available. Good estimations of these measures during the placement process are important. The smallest bounding rectangle enclosing all placed building blocks is normally used to approximate the overall area. An efficient and popular estimation of a wire length is illustrated in Fig. 10.3. The smallest bounding rectangle that encloses all nodes belonging to a signal net is determined. The length of the wire is estimated as one-half the perimeter of the rectangle. If there are critical paths with delays to be minimized, the objective function may assign higher cost factors to them so their effects are appropriately emphasized.

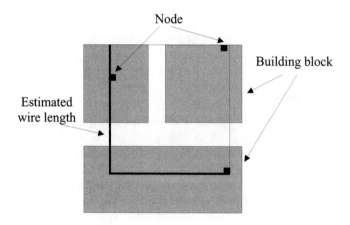

Fig. 10.3 Estimate of wire length.

Constructive Placement

Constructive placement approaches build a feasible placement by starting from a seed building block. The remaining building blocks are chosen one at a time and added to the placement. Often the building blocks are ordered according to their connectivities, defined as the number of connections, with the blocks in the existing placement. The one that has the maximum number of connections with the existing placement is chosen to be placed next. This step is repeated until the placement is complete. We will describe an approach called min-cut that can be used to organize the building blocks for constructive placement.

An example constructive placement procedure is demonstrated in Fig. 10.4. A seed building block is chosen. This selection could be random or based on the connectivity with the I/O pads. We place this seed building block into the lower left corner of the placement area. A cost figure of the existing placement, typically a function of the overall wiring length, is computed. A building block selected from the group to be placed is added to the placement and the cost figure is updated accordingly. The location of the newly added building block is chosen to minimize the cost increase. The placement should grow evenly on its upper, diagonal, and right sides simultaneously.

Fig. 10.4 Constructive floorplanning approach.

Min-Cut Placement

The min-cut algorithm is a graph theory based two-way partitioning algorithm. This algorithm has been applied to placement and is a popular approach for floorplanning. It can be viewed as a refinement of constructive placement described above.

Fig. 10.5 illustrates a min-cut placement process. Given a circuit, an imaginary vertical cutline c_1 is used to partition the building blocks into two groups A and B so that the number of interconnections between these two groups is minimized. The objective is to keep the strongly connected building blocks in the same group so that their interconnection lengths are minimized.

The process is repeated on groups A and B. Group A is partitioned into two groups A_1 and A_2 using a horizontal cutline c_2. Similarly, group B is partitioned into two groups B_1 and B_2 using a horizontal cutline c_3. More partitioning can be introduced until each group contains only one building block. Each partitioning step should attempt to balance the overall areas of the two groups generated.

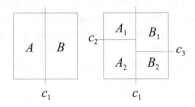

Fig. 10.5 Min-cut placement.

The constructive placement approach described above needs an algorithm to select the building block to be placed next. Often the min-cut algorithm is used to order the building blocks into a linear list in a way so that a line drawn between any consecutive building blocks in the list will cut a minimum number of signal nets. An approach similar to what we described above can be used with only vertical cutlines applied. Fig. 10.6 illustrates such a list in which B_1 to B_n are building blocks ordered by cutlines c_1 to c_{n-1}.

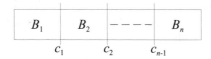

Fig. 10.6 Min-cut linear list for constructive placement.

Iterative Improvement Placement

Iterative improvement is one of the most popular placement methods. The basic iterative improvement approach is characterized by a way of initial solution generation, a way of perturbing an existing placement into a new one, a cost function to evaluate a solution, and a criterion to accept a solution.

An initial solution is created in the first step of this approach. There are many ways to generate an initial placement solution. For example, building blocks can be randomly placed on the chip area. A constructive placement approach can also be used to generate an initial solution.

The objective function of an iterative improvement approach can include any terms that we described above, such as the overall area and overall interconnection length. Notice that overlapping building blocks are usually not allowed in a layout. The objective function could include the overlapping area as a penalty term with a variable weight. This variable weight can be initially set to a small value so that overlapping is temporarily acceptable at the beginning of the process. This will simplify the generation of placement solutions. The algorithm must ensure that building block overlaps are eliminated in the final placement. This goal may be achieved by gradually increasing the penalty weight during the process so that overlaps are completely rejected when the final solution is produced.

A change to a placement can be made by moving a building block or interchanging locations of two building blocks. In some cases, the orientation of a building block can also be changed by rotation and flipping. Usually a random process is used to perturb a placement into another. For example, a pair of blocks may be chosen randomly for exchanging.

Every time a new placement is generated, it is evaluated by computing its cost according to the objective function. There are two possible outcomes. The cost of the new placement may be lower than that of the original one. In this case, the placement is improved. It is accepted and used for further improvements. Alternatively, the cost of the new placement may be unchanged from or even higher than that of the original placement. Traditionally, an iterative improvement approach accepts only solutions that have lowered the cost. Solutions with increased costs are rejected. The original solution is kept and another new solution is generated. Since only cost improving solutions are accepted, this is called a greedy algorithm.

A greedy algorithm has an inherent shortcoming. It is possible that a solution achieved has a cost higher than that of the optimum but no perturbation can cause a further cost reduction. In such a situation, the algorithm is trapped at a local minimum. Fig. 10.7 shows this situation of a greedy algorithm. Since only downhill moves are accepted, the search may end up in a local minimum if no moves can bring it out of the cost function valley.

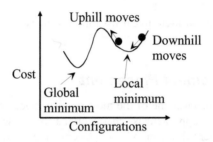

Fig. 10.7 Trapping at a local minimum.

A probabilistic hill climbing technique called simulated annealing has been introduced into the iterative improvement placement. The objective is to escape from local minima. A simulated annealing process is analogous to the process of annealing which attempts to bring a piece of metal into a highly ordered low energy state. In an annealing process, the metal is heated to a high temperature until it is melted into a liquid state. It is then cooled down very slowly. It has been shown in physics that if the initial heating puts the metal into a sufficiently random state, and if the cooling is slow enough to ensure a thermal equilibrium, then the atoms in the metal will move into positions corresponding to a global energy minimum.

Simulated annealing considers the building blocks to be placed as the atoms which are to be moved into a global cost minimum. Usually it starts with a random initial placement. An altered placement is generated. The resulting change in cost, Δc, is calculated. A down-hill move results in $\Delta c < 0$ and is accepted. If $\Delta c \geq 0$, indicating an up-hill move, it is accepted with probability

$$P = e^{-\frac{\Delta c}{t}} \tag{10.1}$$

in which t is the simulated temperature. It can be seen that for the same Δc value, the probability of accepting a cost increasing move is higher at higher temperature t. Simulated annealing decreases the temperature t as it proceeds so that the probability of accepting cost increasing moves decreases. This allows the search to climb out of local minima in search for a global minimum.

Simulated annealing is an important algorithm and is applicable to many combinational optimization problems in VLSI design. The quality of its result depends on the initial value of temperature used and the cooling schedule, which are often experimentally determined. The higher the initial temperature and the slower the cooling schedule the better is the result. The computational time required to generate the solution is proportional to the steps in which the temperature is decreased.

As an example, a simulated annealing process can be set up by choosing an initial temperature t_i such that ~90% of moves are accepted. The decrement of temperature is controlled by

$$t_{i+1} = \alpha t_i \tag{10.2}$$

where t_{i+1} and t_i are the next and current temperatures, respectively, and α is typically a number between 0.8 and 0.95. The annealing process should stay at a temperature long enough so that a sufficient number of moves are tried out to explore the solution space. A rule of thumb is to generate n moves for each building block involved in the placement, where n is 20% to 50% of the total number of blocks to be placed.

Force-Directed Placement

A numerical optimization technique based on the concept of Hooke's law in physics can be used to solve a placement problem. The placement problem is transformed into a set of simultaneous linear equations, which can be solved to determine the ideal coordinates of the building blocks.

According to Hooke's law, if two masses are connected by a spring, the force pulling each other is proportional to the distance separating them. The basic idea behind the force-directed placement is that building blocks connected by a signal net exert forces proportional to their distance on one another. In physics, a spring constant is defined to represent the strength of the spring so that the force with which two masses pull each other is $k \times d$, where k is the spring constant and d is the distance. Similarly, in a force directed placement approach, a weight w_{ij} is assigned to the interconnection between two blocks i and j. For example, a stronger weight indicates more signal nets. Alternatively, a higher weight can be assigned to a more critical signal net. The force of attraction between the blocks is proportional to $w_{ij} \times d_{ij}$, where d_{ij} is the distance between blocks i and j.

If a block is connected to several blocks, it will experience a total force exerted by these blocks. If the block is allowed to move freely, it would move in the direction until the resultant force on it is zero. Based on this principle, numerical technique can be used to solve the optimization problem. A set of simultaneous linear equations can be set up to determine ideal coordinates for the blocks.

When there are a lot of blocks involved, it is less practical to apply the above approach. The force directed method could then be generalized into a constructive placement method. Every time a building block is selected to be added to the existing placement, its zero-force location is computed. Since the ideal location may be occupied by a block previously placed, the process can be further enhanced by performing iterative improvement on the solution obtained.

10.4 Routing

The result of a placement process specifies the exact locations and orientations of building blocks. The next step is to perform a routing process. The objective is to determine the geometric layouts of all signal nets. Traditionally, signal nets must be routed within routing spaces reserved during the placement procedure, which are the regions on the chip area not occupied by the building blocks. With the availability of five or more metal layers, this is no longer the case. If the building blocks were created with two or three metal layers, over the block routing with the remaining metal layers would be possible. Even though the separation of intra-block and inter-block routing layers ensures that no unwanted electrical connections would be formed accidentally, special attention must be paid to issues such as signal coupling.

In some cases, the objective of the routing problem is to minimize the total wire length, while completing all the connections. This requirement can also be enhanced by making sure that each signal net meets its timing budget. Special signal nets such as clocks, and power and ground lines require special treatment. The routing problem is very hard since in a VLSI circuit, there are tens of thousands of nets to be routed and each of them may be routed in hundreds of possible ways.

A typical routing process is carried out in three steps. The routing space is divided into routing regions. These routing regions must meet the requirements of the router to be used. A global router is used to decide a route for each net by specifying roughly which routing regions it should go through. The result of the global routing is to partition a given routing problem into a set of detailed routing problems. A detailed router determines the actual layout of signal nets in each routing region. Wire segments and vias are created to realize the required interconnections. Specialized routers have been designed to route power/ground lines and clock signals.

Shortest Path Problem

All routing algorithms more or less depend on the capability of finding the shortest path between two signal nodes in the routing region. One of the signal nodes is selected as the starting point, from which paths are explored until the destination node is reached. The path exploration is commonly solved as a breadth-first search problem. An example shortest path problem is illustrated in Fig. 10.8.

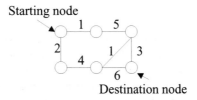

Fig. 10.8 Shortest path problem.

The problem in Fig. 10.8 is to find the shortest path between the starting node and the destination node. The starting node is marked with a cost of zero. Paths are extended in all possible directions from marked nodes to adjacent nodes. Each adjacent node is then marked with the cost (i.e., the weight) associated with the edge connecting the node to the previous node plus the cost of the previous node. The cost of a node can be updated if the new cost is less than the cost already on the node. The marking process continues until all nodes are marked and no further changes are possible. The result of applying this procedure to Fig. 10.8 is shown in Fig. 10.9. The shortest distance between the starting and destination nodes has been determined to be 9.

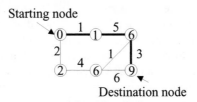

Fig. 10.9 Marking of a shortest path problem.

Upon the completion of the marking, every node in the graph is marked with the lowest cost to it from the starting node. The shortest path is then identified by tracing back from the destination node to the starting node. This can be done by repeatedly finding the neighboring node that is marked with a cost equal to that of the current node minus the cost of the connecting edge. The shortest path is indicated with thick lines in Fig. 10.9. In general, multiple shortest paths may exist between the starting and destination nodes.

Detailed Routing

Detailed routing deals with the determination of wiring paths in a given routing region. A few of the popular detailed routing algorithms are presented before we move on to a discussion of global routing.

Maze Routing

Maze routing is a general-purpose routing algorithm which finds a path between two points in the routing space. The routing region is modeled as a grid graph. A routing surface is considered a collection of square cells arranged into a rectangular array. The size of the square cells is defined according to the design rules so that wires can be routed through adjacent cells without violating the rules for wire width and spacing. Cells that are already occupied by wiring, building blocks, etc. are marked as blocked. Fig. 10.10 shows the grid model of an example routing space. The blocked cells are shaded in the grid model.

Given a grid graph, and a number of interconnecting terminals, the maze routing problem is to find a path connecting these terminals. We describe below a 2-dimensional maze routing algorithm based on the breadth-first search shortest path algorithm. More general maze routing problems have to consider higher dimensional (> 2) grid graphs; however, the techniques we are to describe essentially remain the same.

The Lee algorithm is the most widely used algorithm for finding a path between any two cells on a planar rectangular grid. It is very simple and guarantees the finding of an optimal solution if it exists.

The Lee algorithm can be divided into three phases. The first phase is to perform a breadth-first search. This search can be viewed as a wavefront expanding and propagating from the source (i.e., starting cell). The source is labeled '0' and the wavefront propagates to all the unblocked cells adjacent to the source. Every unblocked cell adjacent to the source is marked with the label '1'. Then every unblocked cell adjacent to cells with the label '1' is marked with the label '2'. In general, every unblocked cell adjacent to cells with the label 'n' is labeled as '(n + 1)'. This wavefront expansion process continues until the target (i.e., destination cell) is reached. If there is no more expansion possible before the target is reached, no route between the source and target is possible. Fig. 10.11 shows the labeling of a wavefront expansion process.

The second phase of the Lee algorithm is to back trace from the target to the source. This can be done by following the labels in descending order from the target created in the wavefront expansion phase. Notice that generally more than one path may exist, but due to the nature of a breadth first search, any of these paths is guaranteed to be the shortest distance path between the source and the target. The shortest path identified by the back tracing step is illustrated in Fig. 10.12.

The Lee algorithm requires a large amount of storage space and its performance degrades rapidly when the size of the grid graph increases. There have been numerous attempts to modify the algorithm to improve its performance and reduce its memory requirements. References to a few of these variations are listed at the end of the chapter.

One variation of the Lee algorithm that is worth mentioning is the creation of a parameter called expansion distance. Recall in the Lee algorithm, only the cells immediately adjacent to a labeled cell is explored at each wavefront expansion step. This is referred to as an expansion distance of one. The adoption of different expansion distances provides maze routing algorithms with different flavors. One of these maze routing algorithms with an expansion distance greater than one is the line-probe routing that we will discuss next.

Line-Probe Routers

Line-probe routers were proposed as an alternative to the Lee algorithm. The routing space is modeled as lists of lines. The basic principle of a line-probe router is to project wires (i.e., line-probes) as far as possible in both the horizontal and vertical directions from both the source and the target. If a line-probe is stopped by some obstacles, the algorithm chooses an escape point along the current line-probe to send additional line-probes out. This continues until the line-probes originated out of the source intersect the line-probes originated from the target. The routing is determined by tracing a sequence of lines which connect the source and the target.

The line-probing procedure can be accomplished by setting an expansion distance of ∞ (infinite) in the maze routing algorithm. In the wavefront expansion phase, a labeled cell is expanded in all 4 directions until an obstacle or another labeled cell is reached. Every grid node on the line segments generated is an "escape" point, from which further expansion is performed. This continues until the target is reached. A backtracing phase determines the path. Fig. 10.13 shows the result of a line-probe routing. Similar to the breadth-first search, the line-probe routing algorithm guarantees to find a path if one exists. However, the route found may not be the shortest one. In contrast, the result route is guaranteed to have a minimum number of turns.

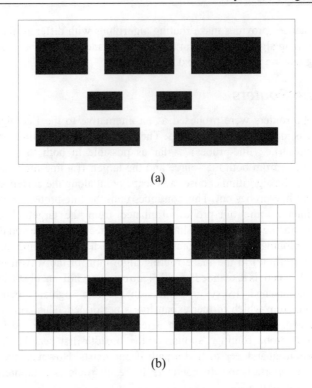

Fig. 10.10 (a) Routing space; (b) Grid graph model for maze routing.

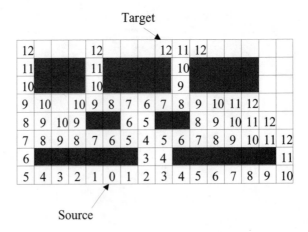

Fig. 10.11 Labeling for the wavefront expansion in the Lee algorithm.

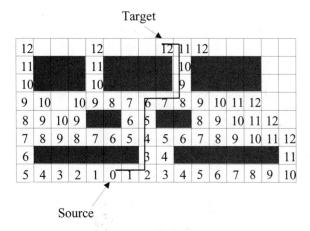

Fig. 10.12 Shortest path found with the Lee algorithm.

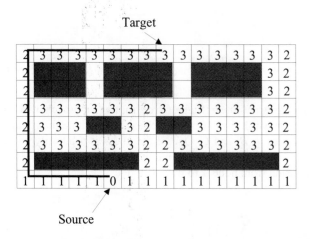

Fig. 10.13 Line-probe maze routing.

While in practice it is rarely used, it is interesting to note that the quality of a route, defined by its distance and number of turns, can be traded off by choosing an expansion distance between 1 and ∞.

Rectilinear Steiner Tree Routing

The routing algorithms presented so far deal with the routing of a signal net between two nodes. A straightforward way to extend the maze routing algorithm to route *n*-terminal nets ($n > 2$) is to decompose them into multiple two-terminal nets.

These two-terminal nets are then routed by a maze routing or line-probe routing algorithm. Fig. 10.14 demonstrates the result of this approach. This simple extension of the maze routing approach may produce a path that is longer than necessary since there is no interaction between the two-terminal nets. The quality of the routing result depends on how the nets are decomposed into subnets and the order of routing of the subnets. Generally suboptimal results are produced.

A slightly more complicated extension of the Lee algorithm can be used to determine the shortest path for an n-terminal net. This is called the Rectilinear Steiner Tree approach. The biggest advantage of this approach is to allow a later route to be connected to any point on a route completed earlier.

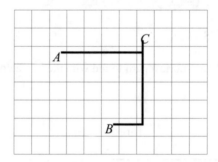

Fig. 10.14 An n-terminal net routed by a maze routing algorithm.

We explain the rectilinear Steiner tree routing approach with the n-terminal routing problem shown in Fig. 10.14, which is to determine the routing of a net that connects three nodes A, B, and C. One of the nodes is arbitrarily chosen to perform the wavefront expansion until a second node is labeled. Fig. 10.15 shows the wavefront expansion from node A to node C.

4	3	2	3	4	5	6			
3	2	1	2	3	4	5	C 6		
2	1	A	1	2	3	4	5	6	
3	2	1	2	3	4	5	6		
4	3	2	3	4	5	6			
5	4	3	4	5	6				
6	5	4	5	6	B				
	6	5	6						

Fig. 10.15 Expansion from node A.

After the second node is identified, a wavefront expansion is performed from the second node to the first node. A modified labeling scheme is applied in this wavefront expansion. The second node (C in our example) is labeled with the label assigned to it in the first wavefront expansion and the wavefront expansion is done with descending labels. Fig. 10.16 shows this descending labeling wavefront expansion from node C to node A.

Now we have two sets of labels, one from the wavefront expansion originated from the first node (A in our example) and one from the wavefront expansion originated from the second node (C in our example). These two sets of labels are superimposed together. Only cells with identical labels from both wavefront expansions retain their labels. The other labels are erased. The cells of which their labels are kept define a restricted routing area between the first (A) and second (C) nodes, which encloses all shortest paths between them. The restricted routing area for the ongoing example is illustrated in Fig. 10.17.

An ascending labeling wavefront expansion is then carried out for the third node (B) until the restricted routing area defined for the first and second nodes is met. The wavefront expansion for node B is shown in Fig. 10.17.

		1	2	3	4	5	4	3	2	
	1	2	3	4	5	C_6	5	4	3	
A	1	2	3	4	5	4	3	2		
		1	2	3	4	3	2	1		
			1	2	3	2	1			
				1	2	1				
		B		1						

Fig. 10.16 Descending labeling wavefront expansion from node C.

					C			
	A			4				
			4	3	4			
			4	3	2	3	4	
		4	3	2	1	2	3	4
	4	3	2	1	B	1	2	3	4
		4	3	2	1	2	3	4

Fig. 10.17 Expansion from node B.

A line-probe routing is then performed from the point where the wavefront expansion of the third node intersects the restricted routing area. Three routes are determined from the intersection to the three nodes, respectively. The result is a Steiner tree routing path. The Steiner tree routing path for the ongoing example is shown in Fig. 10.18.

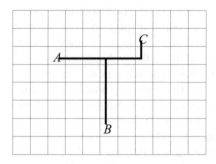

Fig. 10.18 Steiner tree routing path.

Channel Routing

One important group of detailed routing algorithms is channel routers. A channel is a rectangular routing region bounded by two parallel rows of terminals (or pins). The locations of these terminals are fixed along the top and bottom of the channel. The top of a channel is called the top boundary. The bottom of a channel is called the bottom boundary. Each terminal is assigned a number which represents the net to which that terminal belongs to. Vacant terminals are assigned with the number 0. A vacant terminal does not belong to any net and therefore requires no electrical connection. The channel may also have pins in the middle to make connections with higher routing layers.

Unlike the maze routers described above, channel routers consider all nets to be routed in a channel in parallel. Fig. 10.19 shows the elements used in a routing channel. A channel routing problem is specified by the following parameters: channel length which is defined as the horizontal dimension of the routed channel, top and bottom terminal lists, and the number of layers.

A channel routing can be performed with only two routing layers. Generally, vertical segments (branches) of routing paths in a channel are on the same layer. The horizontal segments (trunks) of routing paths in a channel are on the other layer. The lines along which the trunks are placed are called tracks.

The main objective of channel routing is to minimize the channel height, the total number of vias, and to minimize the length of any particular net. Each channel is initially assigned a height by the routing area partitioning process and the channel routing's task is to complete the routing within the assigned space. If it cannot complete the routing in the assigned height, the channel has to expand which makes the initial placement invalid.

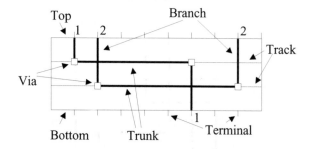

Fig. 10.19 Elements of a routing channel.

Left-Edge Channel Routing Algorithm

The left-edge channel routing algorithm was the first algorithm developed for channel routing. At least two routing layers are used. The algorithm reserves one routing layer for trunks (horizontal wire segments) and another layer for branches (vertical wire segment). This implies that a via has to be used whenever a net changes direction. The basic left-edge algorithm restricts each signal net to have at most one trunk.

The trunks of the signal nets are sorted in ascending order of their left endpoints. The algorithm then allocates tracks to the trunks by considering them one at a time according to their sorted order. An example of routing four signal nets in a channel is shown in Fig. 10.20. The four trunks (A to D) are first sorted according to their left endpoints. After assigning the trunks to the tracks, the branches can be easily added.

In order to fill the track as much as possible, multiple trunks should share a single track if possible. Fig. 10.21 shows an improvement of the above example. The number of tracks in the channel is reduced to two.

Consider the channel routing problem shown in Fig. 10.22. If we switch the track assignments for trunks A and B so that trunk B is on top of trunk A, an unroutable situation occurs. This is called a vertical constraint, which specifies that trunk A must be placed above trunk B. Vertical constraints are usually represented in the form of a directed graph as shown in Fig. 10.23. Another vertical constraint exists between trunks B and C. This vertical constraint forces the use of three tracks to complete the routing. Fig. 10.24 shows the use of a "dogleg" segment to reduce the number of tracks to two. Fig. 10.25 shows that doglegs can also be used to solve circular vertical constraints.

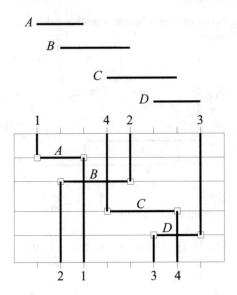

Fig. 10.20 Basic left-edge channel routing.

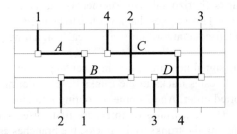

Fig. 10.21 Improved left-edge channel routing.

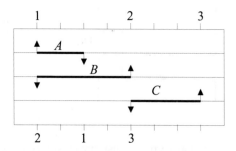

Fig. 10.22 Vertical constraints.

Fig. 10.23 Graphical representation of the vertical constraints in Fig. 10.22.

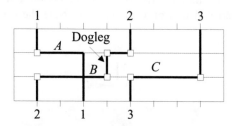

Fig. 10.24 Application of a dogleg.

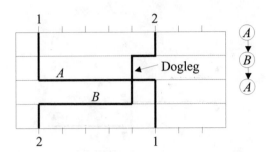

Fig. 10.25 Circular vertical constraints resolved by a dogleg.

Greedy Channel Router

The left-edge algorithm and its variations route the channel one track at a time. The greedy channel router divides a given channel into a number of vertical routing regions called columns. The router attempts to route the channel column by column. This algorithm routes the nets by working across the channel from left to right. The routing within a column is completed before proceeding to the next column. In each column, the router applies a number of "greedy" steps to route the nets in a given column. Doglegs are allowed in any column of the channel.

In the first step, the terminals at the top and bottom of the column are brought to a trunk. Such a connection is made by using the empty track closest to the terminal, or a track that already carries the net. An example of this step is shown in Fig. 10.26. A trunk A is created for the top terminal while the bottom terminal is connected to trunk B which was created in a previous column. Note the use of vias to switch a branch to a trunk.

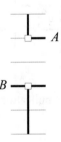

Fig. 10.26 Bringing terminals to the trunks (step 1).

The second step tries to collapse any split signal nets (trunks of the same net present on two different tracks) using a branch. The third step tries to reduce the range of the distance between two tracks of the same net. The fourth step tries to move the nets closer to the destination. An example of the second step is shown in Fig. 10.27.

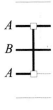

Fig. 10.27 Collapsing split nets (step 2).

Since not all split nets can be collapsed in a column, the third step is thus to reduce the distance between split nets. An example is shown in Fig. 10.28. Step 2 is applied in this example to collapse net A. The vertical segment used to collapse net A blocks the collapsing of net B. Step 3 is thus used to bring the two segments of net B closer together.

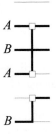

Fig. 10.28 Reducing range of split nets (step 3).

The fourth step attempts to bring all nets closer to their destinations. For example, if the terminal of a net is located on the top of the channel, the router tries to move the net to a higher track. An example is shown in Fig. 10.29.

The fifth step is to create a new track if a terminal failed to enter the channel in step 1. The terminal is then brought to the new track and the routing continues. Due to the greedy nature of this channel router, there is no guarantee on the quality of its routing result. The completion of a routing channel is also not guaranteed.

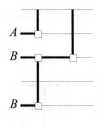

Fig. 10.29 Bringing a net closer to its destination (step 4).

Switchbox Routing Algorithm

The switchbox routing problem is an extension to the channel routing problem. In addition to having terminals on the top and bottom of the routing area as in a channel, a switchbox is a rectangular routing area with terminals on all four sides of the region. It may also have connections in the middle to make connections with upper layers. The greedy channel routing algorithm has been extended to perform switchbox routing. For example, an additional step can be added to the greedy channel router to bring the terminals located on the left side of the switchbox into the first column.

Global Routing

The objective of a global routing process is to divide a large routing problem into a number of smaller, more manageable routing problems. The entire routing space, space that is not occupied by the building blocks and the space above building blocks allowing over-the-block routing, is partitioned into routing regions (e.g., routing channels and routing switchboxes). Each routing region has a limited capacity which is defined by the maximum number of nets that can pass through the region. The objective of global routing is to assign each net to a set of routing areas.

The maze routing algorithm can be adopted to perform a global routing. An example is shown in Fig. 10.30. A graph model is superimposed on the result of a placement process. The routing areas are represented as nodes and adjacent routing areas are connected with edges. The edges have weights that model the length and capacity of the routing areas. The weights can be updated dynamically to reflect the cost of routing a net through it due to congestion. In Fig. 10.30, in addition to performing the routing in the space between building blocks, the routing above one of the building blocks is allowed. A node is placed on top of this building block to reflect this condition. A shortest path algorithm can be used to determine the route of a net.

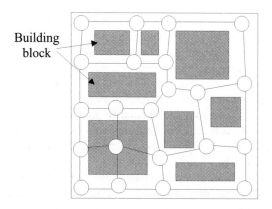

Building
block

Fig. 10.30 Global routing graph.

Ideally, a minimum Steiner tree path should be determined for each net. Unfortunately, this problem is NP-complete. Heuristic techniques have thus been developed to produce acceptable results.

As in the Lee algorithm, nets are sequentially selected for routing. The edge weights are updated after each net is routed. If a net involves more than two terminals, the minimum distance Steiner tree algorithm can be applied. An initial global routing solution is produced after all the signal nets have been routed.

An initial global routing solution can be iteratively improved by repeatedly modifying the current routing solution. A routing solution can be modified by selecting and deleting connections, and reconnecting them using alternate paths. Different approaches have been developed to select and reroute nets. For example, nets in channels with maximum density are rerouted.

10.5 Summary

Physical design deals with the generation of layout information for fabrication. Layout tools are available for the design of library cells and custom designs. A layout generator can be developed to generate parameterized regular layout structure. Tools have been developed to verify that a layout conforms with the design rules, electrical rules, minimum density rules, and antenna rules specified for a given fabrication process.

Placement and routing are two important procedures in the physical design of an integrated circuit. An understanding of place-and-route algorithms allows a designer to use the CAD tools intelligently and obtain better results.

An abstract placement is called floorplanning, which is used to plan the shapes of building blocks. A number of placement algorithms have been discussed. An iterative improvement placement algorithm has the problem of being trapped into a local minimum of its cost function. This is due to the fact that these algorithms only pursue short-term gains. A probabilistic hill-climbing algorithm called simulated annealing was introduced. Simulated annealing attempts to escape from local minima by accepting some cost-increasing solutions in a controlled manner.

In the field of routing, we have presented maze routing algorithms which serve as the foundation of many global and detailed routing algorithms. Channel routing and switchbox routing algorithms have been discussed. The availability of more than two metal layers has increased the flexibility and capability of detailed routing. However, the downside is that the routing complexity is also increased.

No routing algorithm can guarantee a complete routing in a given routing area. When a routing cannot be completed, nets have to be selected to be ripped up and rerouted. The key is to select wires which allow failures to be routed in alternate paths. The placement result may also have to be modified to reduce net congestion in certain routing regions.

10.6 To Probe Further

Physical Design Automation:

Each of these books has excellent coverage for the topics that we have discussed in this chapter. They also provide extensive lists of references for individual subjects in physical design automation.

- N. Sherwani, *Algorithms for VLSI Physical Design Automation*, 3rd *Edition*, Kluwer Academic Publishers, 1999.

- B. T. Preas and M. J. Lorenzetti, ed., *Physical Design Automation of VLSI Systems*, The Benjamin/Cummings Publishing Company, Inc., 1988.

- S. M. Sait and H. Youssef, *VLSI Physical Design Automation Theory and Practice*

Commercial Physical Design Tools:

- http://www.mentor.com
- http://www.cadence.com

Magic Layout Tools:

- http://www.research.digital.com/wrl/projects/magic

Commercial Compaction Tools:

- http://www.sagantec.com

- http://www.rubicad.com

Simulated Annealing:

- S. Kirkpatrick, C. D. Gelatt, Jr., and M. P. Vecchi, "Optimization by simulated annealing," *Science*, vol. 220, no. 4598, 1983, pp. 671-680.

Layout Editor:

- J. K. Ousterhout, G. Hamachi, R. Mayo, W. Scott, and G. Taylor, "Magic: a VLSI layout system," *Proc. 21st Design Automation Conf.*, June 1984, pp. 152-159.

Layout Extractor:

- W. S. Scott and J. K. Ousterhout, "Magic's circuit extractor," *IEEE Design and Test*, 1986, pp. 24-34.

Design Rule Checker:

- G. S. Taylor and J. K. Ousterhout, "Magic's incremental design-rule checker," *Proc. 21st Design Automation Conference*, 1984, pp. 160-165.

Placement:

- C. Sechen and A. L. Sangiovanni-Vincentelli, "TimberWolf3.2: A new standard cell placement and global routing package," *Proc. 23rd Design Automation Conference*, 1986, pp. 142-146.

- M. A. Breuer, "A class of min-cut placement algorithms," *Proc. 14th Design Automation Conference*," October 1977, pp. 284-290.

Routing:

- C. Y. Lee, "An algorithm for path connection and its application," *IRE Trans. on Electronic Computers*, vol. EC-10, 1961.

- K. Mikami and K. Tabuchi, "A computer program for optimal routing of printed circuit connectors," *Proc. IFIPS*, pp. 1475-1478, 1968.

- D. W. Hightower, "A solution to the line routing problem on a continuous plane," *Proc. 6th Design Automation Workshop*, 1969, pp. 1-24.

- D. N. Deutsch, "A 'dogleg' channel router," *Proc. 13th Design Automation Conference*, 1976, pp. 425-433.

- D. Braun, J. L. Burns, F. Romeo, A. Sangiovanni-Vincentelli, and K. Mayaram, "Techniques for multilayer channel routing," *IEEE Trans. on Computer-Aided Design of Integrated Circuits and Systems*, vol. 7, no. 6, June 1988, pp. 698-712.

- R. L. Rivest and C. M. Fiduccia, "A greedy channel router," *Proc. 19th Design Automation Conference*, 1982, pp. 418-424.

- J. Cong, D. F. Wong, and C. L. Liu, "A new approach to three- or four-layer channel routing," *IEEE Trans. on Computer-Aided Design of Integrated Circuits and Systems*, vol. 7, no. 10, October 1988, pp. 1094-1104.

- J. Cong and C. L. Liu, "Over-the-cell channel routing," *IEEE Trans. on Computer-Aided Design of Integrated Circuits and Systems*, vol. 9, no. 4, April 1990.

Compaction:

- W. H. Wolf, R. G. Mathews, J. A. Newkirk, and R. W. Dutton, "Algorithms for optimizing two-dimensional symbolic layout compaction," *IEEE Trans. on Computer Aided Design*, vol. CAD-7, no. 4, April 1988, pp. 451-466.

- W. Shiele, "Improved compaction with minimized wire length," *Proc. 20th Design Automation Conference*, 1984, pp. 166-172.

- S. Sastry and A. Parker, "The complexity of two-dimensional compaction of VLSI layouts," *Proc. IEEE International Conf. on Circuits and Computers*, 1982, pp. 402-406.

10.7 Problems

10.1   Use the basic left-edge channel routing algorithm to route the channel given in Fig. 10.31.

$$\overline{2\ 1\ 2\ 1\ 5\ 3\ 6}$$

$$\overline{5\ 3\ 6\ 4\ 0\ 2\ 4}$$

Fig. 10.31 Channel routing problem for Problem 10.1.

10.2   Repeat Problem 10.1 with the improved left-edge channel routing algorithm. Improve the routing result by introducing dog-legs.

10.3   Route the switchbox given in Fig. 10.32.

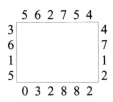

Fig. 10.32 Switchbox routing problem for Problem 10.3.

10.4   Write a program to implement the constructive placement algorithm.

10.5   Write a program to implement the iterative improvement placement algorithm.

10.6   Write a program to implement the simulated annealing placement algorithm.

10.7   Consider a standard-cell library that has a standard height of 5 units. The library provides an inverter (width: 2 units), a 2-input NAND gate (3 units), a 3-input NAND gate (4 units), a 2-input NOR gate (3 units), a 3-input NOR gate (4 units), a 2-input XOR gate (5 units), and a 3-input XOR gate (6 units). Use place-and-route technique to generate a layout diagram for a full adder implemented with these standard-cells.

Chapter 11 Parallel Structures

Two brains are better than one ...

Pipelining and parallel processing can be used to improve system throughputs, often significantly. If an application either processes a large amount of data or contains a large number of iterations (i.e., loops), it is a candidate for parallelism exploration. In a parallel processing system, processing elements cooperate to produce the desired result. This requires the creation of a parallel algorithm which, in many cases, involves the parallelization of an existing sequential algorithm.

Parallel architectures and algorithms for general-purpose parallel processing have been extensively studied for several decades. We are interested in the creation and application of application specific parallel processing techniques enabled by the low cost and high density of VLSI circuits.

General-purpose parallel processing is discussed in this chapter to provide the necessary background for the VLSI parallel algorithm/architecture development to be presented in Chapter 12. There is a main difference between general-purpose parallel processing and special-purpose parallel architectures. With only a few exceptions, general-purpose parallel architectures employ only a few relatively powerful processing elements (PEs) while special-purpose parallel architectures have the potential of using a large number of simple PEs. We use the term "PE" instead of "processor" since the latter often implies a microprocessor. In the realm of VLSI, a PE could be as simple as an integrated multiplier and accumulator.

11.1 Parallel Architectures

An informal definition for the term "parallel processing" is that multiple PEs are utilized in a coordinated manner to support the solving of a problem. The complexity of a PE in a parallel architecture can range from a standalone computer, as in the case of distributed computing, to a much simpler functional unit such as adders and multipliers, as in the case of an application specific parallel processing system.

Ideally, if a task takes n minutes to complete with a single PE, its completion time should be reduced to 1 minute if n PEs are used. This idealistic target is often non-reachable. The jeopardizing factors include the overhead involved in the control and interaction of these PEs and data dependency. Not all operations in an algorithm can be executed in parallel. A strictly sequential algorithm cannot be parallelized.

Consider an operation as simple as adding 10 pairs of numbers. This operation is described in the following pseudo code.

```
for (i=1; i<=10; i++){
    c[i]=a[i]+b[i];
```

}

The above pseudo code describes a loop, in each iteration of which two numbers ($a[i]$ and $b[i]$) are added to produce a sum ($c[i]$). Since there is no dependency between these iterations, the loop can be unrolled into ten independent operations:

```
c[1]=a[1]+b[1];
c[2]=a[2]+b[2];
...
c[10]=a[10]+b[10];
```

Assuming that all $a[i]$'s and $b[i]$'s are simultaneously available, these 10 operations can be performed in 10 individual PEs. The time to obtain all $c[i]$'s is then reduced to that of a single addition plus the overhead. The benefit of parallel processing is evident.

Consider that the problem is now modified into an accumulation of all ten sums, as described in the following pseudo code:

```
c[0]=0;
for (i=1; i<=10; i++) {
    c[i]=c[i-1]+a[i]+b[i];
}
```

This operation shows a dependency between two iterations of the loop. If we follow the pseudo code directly, the computation of $c[i]$ needs $c[i-1]$ computed at an earlier time. However, a little inspection should reveal that it is still possible to explore parallelism in this case, although the application of parallel processing to this operation is less straightforward. This example demonstrates an important fact. The parallelism of a problem may be hidden by a sequential algorithm. It is important to understand the nature of computations in the specific application domain.

Parallel processing, also called multi-processing in some context, should not be confused with the term multi-tasking. A system with a single PE can be seemingly working on multiple processes at the same time. This is done by an approach called time-sharing, in which the PE works on the processes, one at a time, and switches from one to another after a time slice has expired. Another related concept is multi-threading. Each process in a multi-processing operation has its own memory space. In contrast, the threads in a multi-threading operation share the same memory space.

The PEs in these general-purpose parallel architectures are often limited to a few complete central processing units (CPUs). Traditionally general-purpose parallel processing architectures are classified by whether or not the PEs are synchronized to execute the same instruction concurrently and the number of data streams processed by the PEs. Normally the PEs, regardless whether there are single or multiple instruction streams, operate on different data.

A general-purpose computer has a single PE that operates on one datum stream at a time. We refer to this architecture as an SISD (single instruction stream single datum stream) architecture. This architecture is obviously not a parallel structure. We present this architecture here only for the purpose of reference. The speed-up of a parallel architecture is often computed by comparing its performance to that of an SISD architecture. Fig. 11.1 presents the architecture of an SISD computer. The memory element stores instructions, operands, and results.

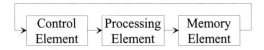

Fig. 11.1 Conceptual view of an SISD architecture.

Even a strictly sequential architecture can be improved by embedding a special form of parallel processing called pipelining. For example, microprocessors can use a pipeline to streamline the multiple phases of a CPU cycle (e.g., fetch instruction, decode instruction, fetch operand, execution, store data, etc.) Pipelining borrows the idea from an assembly line. The principle is to divide the task into a series of sequential subtasks, each of which is executed in a hardware stage that operates concurrently with other stages in the pipeline. Assuming a continuous stream of tasks, a pipeline can improve the throughput of a system by overlapping the subtasks of multiple tasks. Pipelining is an important technique in VLSI architecture and we will discuss its details in this chapter.

According to the taxonomy that uses instruction and data streams to classify parallel architectures, four architecture classes are possible. We have already explained the SISD architecture. Two other practical parallel architectures are SIMD (single instruction stream multiple datum stream) and MIMD (multiple instruction stream multiple datum stream). A general diagram of a parallel architecture is shown in Fig. 11.2. A number of PEs are interconnected through an interconnection network for the purpose of communication. Typically a parallel architecture operates as a back-end co-processor and its interface to the world is handled by a host processor. In Fig. 11.2 the host processor is shown to communicate with the PEs through an interconnection network.

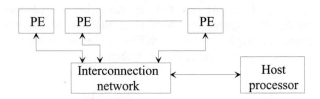

Fig. 11.2   Conceptual view of a parallel architecture.

Due to the complexity of an MIMD architecture, most general-purpose parallel computers have adopted the SIMD architecture. Recently, expensive general-purpose parallel computers have been largely replaced by an approach called distributed computing, which uses a group of networked computers to form an MIMD architecture. While it is theoretically possible to configure an MISD (multiple instruction stream single datum stream) architecture, its usefulness is extremely limited.

Many derivations from the parallel architecture shown in Fig. 11.2 are possible. The architecture can be homogeneous or heterogeneous. A homogeneous architecture contains identical PEs. A heterogeneous architecture employs PEs of different capabilities. The interconnection network can be in the form of a time-shared common bus, a switching network, or a shared memory. The control of the PEs can be centralized or distributed.

General-purpose parallel architectures are especially efficient for vector and matrix operations. They are also capable to be programmed to solve partial differential equations, perform sorting, and do fast Fourier transform (FFT). The performance of a parallel architecture is often measured by its speed-up which is defined as:

$$\text{Speed-up} = \frac{\text{Execution time on a sequential (SISD) architecture}}{\text{Execution time on a parallel architecture}} \qquad (11.1)$$

A parallel architecture with n identical PEs is at most n times faster than its single PE sequential counterpart. Practically the speed-up is usually much less than this. The performance of a parallel architecture is hampered by its operating overhead and data dependency. The overhead of a parallel architecture involves the control for inter-PE communication, data distribution, and result collection.

A rule of thumb can be used to judge whether it is worth the cost of using a parallel architecture. Computer scientists have proposed that the speed-up of a practical parallel architecture comprising n PEs should be at least $\log_2(n)$.

11.2 Interconnection Networks

As shown in Fig. 11.2, one essential element is the interconnection network, which connects the PEs together in a parallel architecture. Ideally, the PEs should form a fully connected graph, in which each PE is connected to any other processor. Fig. 11.3 shows two fully-connected parallel structures.

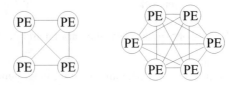

Fig. 11.3 Fully-connected parallel structures.

The fully-connected parallel structure allows any two PEs to communicate directly; so it is the most powerful interconnection network. However, a fully-connected interconnection network with n PEs requires $n \times (n - 1)$ links, which apparently becomes expensive with a large n.

A fully-connected interconnection network can be implemented with a cross-bar interconnection network. The structure of a 4-node cross-bar interconnection network is shown in Fig. 11.4. The circles at the intersections between the vertical and horizontal lines in the cross-bar switch are switches, which can be turned on to connect these lines.

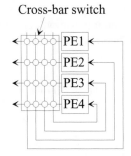

Fig. 11.4 Cross-bar switch implementing a fully-connected interconnection network.

Interconnection networks for parallel architectures have been extensively studied. There are two classes of interconnection networks, static and dynamic. A static network provides fixed and unchangeable connections between PEs. In contrast, dynamic interconnection networks are reconfigurable by setting a set of switch elements. Several popular interconnection network topologies are presented below.

Linear and Ring Interconnection Networks

Fig. 11.5 shows a linear interconnection network with n PEs, in which PEs are organized in ascending order from PE_0 to PE_{n-1}. The connectivity of an interconnection network can be described by a set of routing functions. The linear interconnection network has the following routing function:

$$r_{+1}(i) = i + 1, \text{ for } i = 0 \text{ to } n - 2$$
$$r_{-1}(i) = i - 1, \text{ for } i = 0 \text{ to } n - 1 \tag{11.2}$$

Routing function $r_{+1}(i)$ defines a forward communication from PE_i to PE_{i+1}, which is undefined for PE_n. Routing function $r_{-1}(i)$ defines a backward communication from PE_i to PE_{i-1}, which is undefined for PE_0. The benefit of this topology is in its low hardware cost. However, communication may have to pass through a number of PEs in order to reach the destination. This topology often results in long communication delays. The longest communication delay occurs between PE_0 and PE_{n-1}, which has to go through $n-1$ PEs.

Fig. 11.5 Linear interconnection network with n PEs.

A ring interconnection network can be formed by connecting the first PE (PE_0) and the last PE (PE_{n-1}) in a linear interconnection network as shown in Fig. 11.6. The ring structure removes the long communication delay between the first PE (PE_0) and the last PE (PE_{n-1}). The ring interconnection network has the following routing functions:

$$r_{+1}(i) = (i+1) \bmod n$$
$$r_{-1}(i) = (i-1) \bmod n \tag{11.3}$$

where the "mod" operator is defined as

$$d \bmod n = d - n \times \left\lfloor \frac{d}{n} \right\rfloor \tag{11.4}$$

Fig. 11.6 Ring interconnection network with n PEs.

Shuffle Exchange Interconnection Networks

A shuffle exchange interconnection network is used to connect n ($n = 2^k$, k is an integer) PEs together. Two routing functions are defined for a shuffle exchange interconnection network: a shuffle function (r_s) and an exchange function (r_e). The shuffle routing function gets its name from the shuffling of a deck of cards. The card deck is evenly divided into two halves. A new deck is then formed by interleaving the cards from the two halves. The shuffle operations of 8 PEs are shown in Fig. 11.7. The shuffle routing function is expressed as

$$r_s(i) = \begin{cases} 2i & \text{for } 0 \leq i \leq \dfrac{n}{2} - 1 \\ 2i+1-n & \text{for } \dfrac{n}{2} \leq i \leq n \end{cases} \tag{11.5}$$

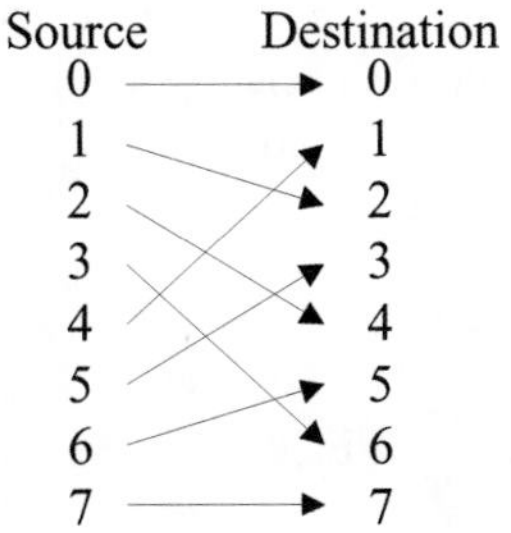

Source   Destination
  0 ──────────────► 0
  1 ────────────◄   1
  2 ────────────▲   2
  3 ────────────▼   3
  4 ────────────▲   4
  5 ────────────▼   5
  6 ────────────◄   6
  7 ──────────────► 7

Fig. 11.7 Shuffle operations of 8 PEs.

As shown in Fig. 11.8, the exchange operations allow two neighbor PEs to exchange information. The exchange routing function is expressed as

$$r_e(i) = \begin{cases} i+1, & i = 0, 2, 4, \ldots, n-2 \\ i-1, & i = 1, 3, 5, \ldots, n-1 \end{cases} \tag{11.6}$$

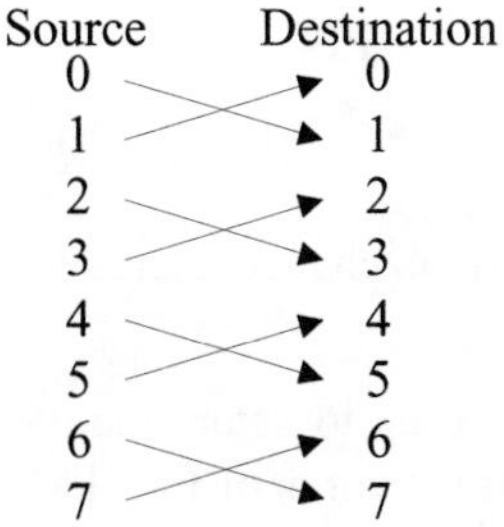

Source   Destination
  0 ────────────► 0
  1 ────────────► 1
  2 ────────────► 2
  3 ────────────► 3
  4 ────────────► 4
  5 ────────────► 5
  6 ────────────► 6
  7 ────────────► 7

Fig. 11.8 Exchange operations.

If the PEs are labeled with binary numbers, the routing functions in (11.5) and (11.6) can also be written as

$$r_s(i) = a_{n-2}...a_1a_0a_{n-1}$$
$$r_e(i) = a_{n-1}...a_1\overline{a}_0$$
(11.7)

where $(i)_{10} = (a_{n-1}a_{n-2}...a_1a_0)_2$.

In (11.7), the shuffle routing function $r_s(i)$ performs a left rotation on the binary label of PE_i to produce its destination. The exchange routing function $r_e(i)$ complements the least significant bit a_0 in the binary label of PE_i to produce its destination. Fig. 11.9 shows a shuffle-exchange interconnection network with 8 PEs. Shuffle connections are represented by solid arrow lines and exchange connections are shown by solid lines.

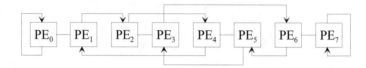

Fig. 11.9 Shuffle-exchange interconnection network with 8 PEs.

The shuffle-exchange interconnection network can be enhanced by changing the shuffle connections (solid arrow lines in Fig. 11.9) into bidirectional links. This can be done by defining unshuffle connections, which are created by reversing the directions of the shuffle connections (i.e., rotating right the binary label of a PE). The result of this enhancement is shown in Fig. 11.10.

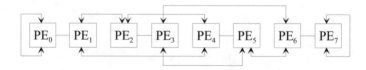

Fig. 11.10 Shuffle-unshuffle-exchange interconnection network with 8 PEs.

A simple algorithm can be used to determine the communication path between two PE's in the interconnection networks of Fig. 11.9 and Fig. 11.10 (see Problems 11.9 and 11.10).

Mesh Interconnection Networks

In a two-dimensional mesh interconnection network, each PE is connected to its north, south, east, and west neighbors. The connections between n PEs are done according to the following routing functions:

$$r_{+1}(i) = (i+1) \bmod n$$
$$r_{-1}(i) = (i-1) \bmod n$$
$$r_{+k}(i) = (i+k) \bmod n, \ k = \sqrt{n} \tag{11.8}$$
$$r_{-k}(i) = (i-k) \bmod n$$

There are no general rules on how the boundary and corner PEs are connected. The routing functions in (11.8) provide wrap-around connections for the boundary and corner PEs. Fig. 11.11 shows a mesh interconnection network with 16 PEs. Wrap-around connections are shown in this interconnection network to fully implement the routing functions in (11.8). An inspection on the routing pattern shows that the worst case inter-PE communication delay is $(k-1)$ PEs.

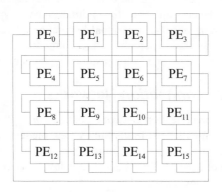

Fig. 11.11 Mesh interconnection network with 16 PEs.

Hypercube Interconnection Networks

A hypercube interconnection network has n PEs, where $n = 2^q$ for some integer $q \geq 0$. Each PE is connected to exactly q neighbors. Fig. 11.12 shows two hypercube interconnection networks with 4 PEs and 8 PEs, respectively. Notice the PEs are labeled with binary numbers (e.g., $PE_{110} = PE_6$) to demonstrate the hypercube routing functions:

$$r_i(a_{q-1}...a_i...a_0) = a_{q-1}...\overline{a}_i...a_0, \ 0 \leq i \leq (q-1) \tag{11.9}$$

There are q routing functions in (11.9), which implies that a PE is linked to q neighbors. Two PEs are connected in a hypercube interconnection network if their binary labels differ by exactly one bit.

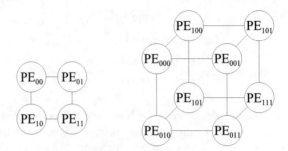

Fig. 11.12 Hypercube interconnection networks.

A q-dimensional hypercube can be built by connecting corresponding PEs (i.e., PEs with identical labels) in two $(q-1)$-dimensional hypercubes. Fig. 11.13 demonstrates the forming of a 4-dimensional hypercube interconnection network by connecting two 3-dimensional hypercubes. PE_{000} in the first 3-dimensional hypercube is connected to PE_{000} of the second 3-dimensional hypercube, and so on.

Binary Tree Interconnection Networks

A binary tree interconnection network connects $n = 2^k-1$ PEs by forming a complete binary tree with k levels. In the binary tree, the levels are labeled from 0 to $k-1$. There are 2^i PEs in level i of the binary tree. Fig. 11.14 shows a binary tree interconnection network for 15 PEs. Each PE at level i is connected to one PE at level $i-1$ and two PEs at level $i+1$.

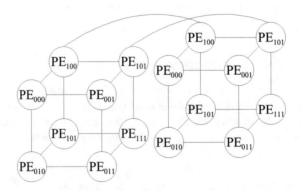

Fig. 11.13 Procedure of forming a 4-dimensional hypercube interconnection network by connecting two 3-dimensional hypercubes (not all connections shown).

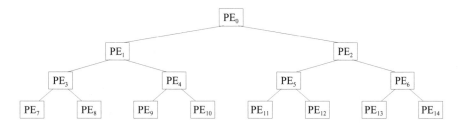

Fig. 11.14 Binary tree interconnection network with 15 PEs.

The routing function of a binary tree interconnection network is left as an exercise for the reader (see Problem 11.11).

Dynamic Interconnection Networks

Earlier in this chapter we introduced the implementation of a fully-connected interconnection network with a cross-bar switch. A cross-bar switch is an example of a dynamic interconnection network since the connectivity between PEs can be changed dynamically. This is in contrast to the static interconnection networks we described above, which provide fixed connections between PEs. Most dynamic interconnection networks were originally proposed to connect between processors and memories in a general-purpose parallel architecture. This need does not appear frequently in special-purpose parallel structures, the subject of Chapter 12.

For the completeness of this introduction, we present the omega interconnection network as an example of a dynamic interconnection network. Given n inputs, the omega interconnection provides $\log_2 n$ stages of switch boxes. Each stage contains $n/2$ switch boxes. Each switch box can be configured into one of the four states illustrated in Fig. 11.15. In addition to providing one-to-one communications, the omega network also supports broadcasting messages by setting certain switch boxes to perform in the upper or lower broadcast mode.

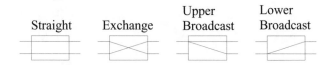

Fig. 11.15 Operating modes of switch boxes for the omega network.

Fig. 11.16 shows the omega interconnection network for 8 inputs (i.e., $n = 8$). A perfect shuffle interconnection is used to connect two adjacent switch box stages. Each switch box is controlled individually. Fig. 11.17 illustrates the connection established between P_0 and M_5.

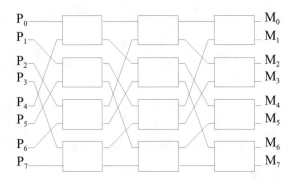

Fig. 11.16 Omega interconnection network with 8 inputs.

11.3 Pipelining

Parallel processing in the form of replicating PEs cannot help a strictly sequential application. However, if the application has a continuous stream of data sets to be processed, performance can still be gained by overlapping the processing of these data sets. This concept has been generalized into an approach called pipelining that explores temporal parallelism.

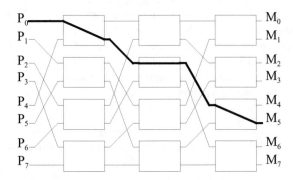

Fig. 11.17 Message routing in an omega interconnection network.

A pipeline architecture works in the same principle as an assembly line to speed up a strictly sequential task. An assembly line arranges the products being produced to pass consecutively through workstations consisting of workers and equipment so that different parts can be added on until completed. In 1913, Ford Motor Company introduced the concept of assembly line into car production and the rate of car producing was increased 8 times.

Fig. 11.18 shows a PE processing a continuous stream of tasks. For example, the PE may evaluate a mathematical expression and the tasks represent different argument sets to be used in the expression. This is analogous to the Henry Ford assembly line that produced multiple identical cars.

Fig. 11.18 Continuous stream of tasks processed by a single PE.

Assume that the steps performed in one task are strictly sequential and thus spatial parallelism among the steps is out of the question. Instead, we can explore parallelism between tasks by replicating the PE. Fig. 11.19 shows that the PE in Fig. 11.18 has been replicated four times.

It is easy to conclude that n PEs will produce a speed-up of n. As we have previously described, this represents a theoretical upper bound of the speed-up and can only be achieved in an ideal case. A number of issues have to be considered in the calculation of a more realistic speed-up. Recall that the tasks arrive in a sequence. A multiplexing scheme is thus needed to direct an arriving task to the next available PE. Similarly, the outputs of the PEs must be directed appropriately. Another factor that may affect the effectiveness of this architecture is the arrival rate of the tasks. The speed-up of n was calculated under the assumption that a task is always ready to enter the system when a PE frees up.

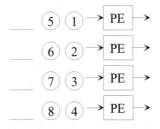

Fig. 11.19 Exploring spatial parallelism between tasks by replicating PEs.

Pipelining provides a less expensive approach to the same problem, which explores temporal parallelism within a PE. Assume that the function performed by the PE has n steps. As shown in Fig. 11.20, the PE is partitioned into n processing stages ($S_0 - S_{n-1}$). Each stage corresponds to a computing step in the mathematical expression. These processing stages perform their operations concurrently and independently. In order to avoid the stages from interfering with each other, the output of a stage is latched by a register, which serves as the input of the following stage. Once its output is latched, the stage can then be released to work on the next task. This type of pipeline architecture that does not have feedback paths is called a linear pipeline.

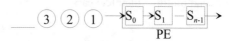

Fig. 11.20 Linearly pipelined PE.

The performance of a pipeline architecture is measured by its throughput, which is determined by its operating frequency. All registers installed between stages are synchronized with the same clock signal. The cycle time of this clock determines the maximum amount of time a stage is allowed to process a task. Ideally all stages would have identical response times. The clock frequency can then be set accordingly so that the registers can correctly capture the stage outputs. Only in ideal cases can we divide the delay time of an operation evenly into stages. In practice the slowest stage becomes the bottleneck of the pipeline since it determines the clock frequency. The following expression is provided for the clock frequency f.

$$f = \frac{1}{t_{max} + t_l} \tag{11.10}$$

where t_{max} is the longest stage delay time in the pipeline and t_l is the delay time of the register.

Example 11.1

A CPU cycle is typically made up of four phases, instruction fetching, instruction decoding, operand fetching, and execution. Consider a CPU that has a cycle time of $4T$. Such a four-phase CPU is shown in Fig. 11.21.

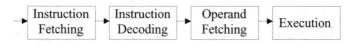

Fig. 11.21 Four-phase CPU.

Assuming that each of the four phases can be individually implemented in a stage with a delay of T, the CPU can be formed as a four-stage pipeline. Registers are inserted between stages to capture intermediate results. We can analyze the operation of such a CPU when it is used to carry out n instructions. This analysis is most conveniently illustrated in the space-time diagrams shown in Fig. 11.22 and Fig. 11.23. Fig. 11.22 shows the space-time diagram for the non-pipelined CPU. Fig. 11.23 shows the space-time diagram for a pipelined CPU. In a space-time diagram, the vertical and horizontal axes represent the stages (space) and the time, respectively. The location where an instruction is being processed at any time is shown by the instruction number marking the intersection of a stage and a time unit. The first instruction (1) enters the CPU at the instruction fetch stage, travels from stage to stage, and finally exits the PE at the execution stage. The whole processing time for the first task is $4T$.

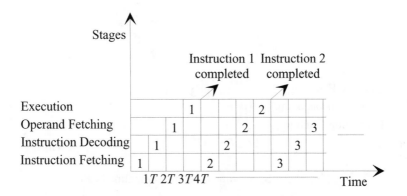

Fig. 11.22 Space-time diagram of non-pipelined CPU.

In Fig. 11.22, after the first instruction has left the CPU, the second instruction enters the CPU. This continues until all instructions are completed. It thus takes $4nT$ time to finish n instructions.

The same space-time diagram can be used to analyze the pipelined CPU. First consider the processing of a single instruction. It takes $4T$ time to go through the CPU, which is the same result as in the original non-pipelined CPU. In fact, the above result has ignored the fact that the delays of the latches have to be added to the overall time. Let us assume that the latch time is much smaller than the major delay time so we can simplify our analysis.

When there is more than one instruction to be processed, the advantage of a pipelined CPU becomes apparent. When the result of the instruction fetch stage has been latched, the instruction stage is free and can be used to process the second instruction. At this time, the first instruction is being processed by the instruction decoding stage. We are thus overlapping the processing of instructions 1 and 2. Similarly, when instruction 1 is being processed by the operand fetching stage, instruction 2 has been advanced to the instruction decoding stage, and a new instruction (3) is processed by the instruction fetching stage. This is shown in the space-time diagram of Fig. 11.23.

The pipelined CPU did not change the fact that an instruction takes $4T$ to complete. Indeed, the first instruction exits the pipelined CPU at the end of $4T$. We call the production time of the first instruction the filling up of the pipeline. On the other hand, the second instruction takes only one additional T to emerge from stage

4. In fact, every following instruction takes an additional T to complete. For the processing of n instructions, the total completion time is then $4T + (n - 1)T$.

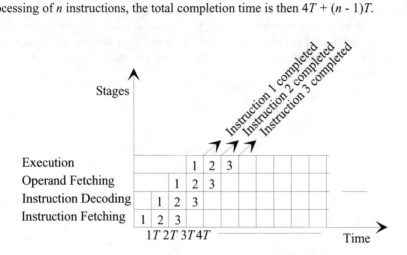

Fig. 11.23 Space-time diagram of pipelined CPU.

In general, if a PE can be divided into k stages, each of which has a delay time of T, the time required to complete n tasks can be expressed as $kT + (n-1)T$. The first term kT is the time required to fill up the pipeline. After a pipeline is filled up, a new result will be produced every clock cycle. Note that if n is much larger than k, the above expression approaches nT. The speed-up of such an idealistic k-stage pipeline PE is the theoretical upper bound of speed-up:

$$\frac{nk}{k+n-1} \approx k \text{ for } n \gg k \tag{11.11}$$

The throughput of a pipeline PE is the number of tasks it can process in a unit time. This is expressed as

$$\text{Throughput} = \frac{n}{(k+n-1)T} \tag{11.12}$$

11.4 Pipeline Scheduling

A pipeline's performance depends on the availability of data. The throughput of a pipeline can only be maintained if we can keep its stages busy. If the data availability does not keep up with the operating speed, the pipeline will have some idle time. One way to improve the hardware utilization rate of a pipeline is to reuse a stage in the processing of a task. Consider the system in Fig. 11.24, in which stage S1 and S4 perform identical functions. This pipeline can be modified to use only a single stage for both S1 and S4. As shown in Fig. 11.25, the output of stage S3 is fed back to the input of stage S1 so that stage S1 can also cover the function originally provided by stage S4. Note that this non-linear pipeline implementation requires the use of appropriate multiplexing and demultiplexing to direct the flow of data stream (see Problem 11.12).

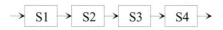

Fig. 11.24 Four-stage pipeline.

Fig. 11.25 Non-linear implementation of the pipeline in Fig. 11.24.

A non-linear pipeline with feedback paths has an inherent limitation since each stage can only process one task at a time. If two or more tasks arrive at the same stage at the same time, we have a "collision" between these two tasks. A non-linear pipeline can thus be considered as a trade-off between hardware cost and performance. We present below an approach that schedules the feeding of a non-linear pipeline for an optimal throughput.

A reservation table that resembles a space-time diagram can be used to analyze a non-linear pipeline for the purpose of scheduling. The reservation table of the non-linear pipeline in Fig. 11.25 is shown in Fig. 11.26. In a reservation table, each row represents a pipeline stage and each column represents a clock cycle. The total number of columns in a reservation table is thus the number of clock cycles to process a task. A mark (X) in the reservation table indicates that the stage is occupied by the task at a specific clock cycle.

Fig. 11.26 Reservation table of the non-linear pipeline in Fig. 11.25.

Apparently, the reservation table of a linear pipeline would have one and only one mark in each row. In a non-linear pipeline reservation table, a task can appear more than once in the same row. This implies that if the initialization of new tasks is not planned carefully, more than one task may require the same stage at the same time. This type of task congestion is called a collision. The objective of scheduling is to determine the scheduling of jobs so that the pipeline is optimally utilized.

The latency of a pipeline is defined as the number of clock cycles between two task initiations. For example, the latency in a linear pipeline is 1. A non-linear pipeline normally cannot maintain unity latency. There may also be different latencies between initiations. A collection of latencies is called a latency sequence. A latency sequence that repeats itself is called a latency cycle. A latency that causes a collision is called a forbidden latency. All forbidden latencies of a non-linear pipeline can be easily obtained from its reservation table. For example, according to the reservation table in Fig. 11.26, the forbidden latency is 3. This means that a new task that is initiated three clock cycles after the last job causes a collision. In general, the forbidden latencies of a reservation table equal to the column distances between two marks in the same row.

A forbidden list can be used to describe all forbidden latencies in a reservation table. This can be expressed by a list $F = \{l_1, l_2, ..., l_r\}$ that contains the column distances l_i between all possible pairs of marks on each row of the reservation table. The forbidden list of the reservation table in Fig. 11.26 is $F = \{3\}$. A collision vector is a binary vector $C = c_n...c_2c_1$, in which $c_i = 1$ if $i \in F$ and $c_i = 0$ otherwise, and n is the longest forbidden latency. The reservation table in Fig. 11.26 has a collision vector $C = 100$.

An n-bit shift register can be used to implement a collision free scheduling for a non-linear pipeline. Initially the shift register is loaded with the collision vector of the pipeline being considered. The shift register is shifted right one bit every clock cycle of the pipeline. A '0' is inserted to the most significant bit of the shift register in each shifting. Every time a new task is initiated, the value stored in the shift register is bitwisely OR'ed with the original collision vector. The result vector of this operation, which describes the current state of the pipeline, is stored back into the shift register. A collision free task initiation is allowed if and only if a '0' is shifted out of the register, which can be used as a control signal to initiate a new task.

This way of scheduling is called a greedy scheduling since a task is initiated as soon as it is possible. This does not generally produce an optimal throughput. For a better throughput, an analysis can be performed by creating a task-scheduling diagram to explore all possible job scheduling of the pipeline being analyzed. This is done by repeating the shift-register operations described above until no more new vector can be generated. The task scheduling diagram for the reservation table in Fig. 11.26 is shown in Fig. 11.27. The "bubbles" in the diagram represent valid values that the shift register may assume. The links connecting two bubbles are labeled with allowable latencies. Notice that a latency of 4 or larger always cause the shift register to resume the original collision vector.

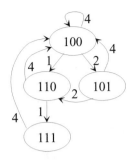

Fig. 11.27 Task scheduling diagram for the reservation table in Fig. 11.26.

Various latency cycles can be identified from a task scheduling diagram. In our ongoing non-linear pipeline example, the following latency cycles can be identified: (1, 1, 4), (1, 4), (2, 2, 4), (2, 2, 1, 4), (2, 4), and (4). An average latency of a latency cycle can be used to determine a scheduling strategy. Since the throughput of a pipeline is inversely proportional to its average latency, the maximum throughput is achieved by choosing a latency cycle that has the minimum average latency. In our example, the latency cycle of (1, 1, 4) is chosen for a minimum average latency of 2.

The scheduling principle that we described above can be generalized to schedule multiple different tasks in a pipeline, each of which has its own reservation table.

11.5 Parallel Algorithms

We present a few examples in this section to demonstrate the utilization of general-purpose parallel architectures. The first example is the multiplication of two matrices, which is a popular application for parallel architectures. Matrix multiplication is important since it is the basic function needed by many scientific computations. Besides multiplication, other important matrix operations include matrix inversion, L-U decomposition, etc. The second example is the sorting of a series of numbers, which is another essential basic computational operation.

In order to appreciate the difference between developing an algorithm for a sequential architecture and a parallel architecture, we begin by describing a sequential algorithm to multiply two matrices.

Matrix Multiplication

Assume the two matrices to be multiplied are

$$[A]_{n \times n} = \begin{bmatrix} a_{11} & a_{12} & \cdots & a_{1n} \\ a_{21} & a_{22} & \cdots & a_{2n} \\ \vdots & & \ddots & \\ a_{n1} & \cdots & \cdots & a_{nn} \end{bmatrix} \text{ and } [B]_{n \times n} = \begin{bmatrix} b_{11} & b_{12} & \cdots & b_{1n} \\ b_{21} & b_{22} & \cdots & b_{2n} \\ \vdots & & \ddots & \\ b_{n1} & \cdots & \cdots & b_{nn} \end{bmatrix} \tag{11.13}$$

We represent a general element in matrix $[A]$ as a_{ik} in which i and k are the row and column numbers, respectively. Similarly, b_{kj} represents a general element in matrix $[B]$ where k and j are the row and column numbers, respectively. The task is to perform the matrix multiplication $[A] \times [B]$ to produce a product matrix $[C]_{n \times n}$. The matrix multiplication can be described with the familiar expression for computing the elements c_{ij} of product matrix $[C]$:

$$c_{ij} = \sum_{k=1}^{n} a_{ik} \times b_{kj}, \ (1 \le i, j \le n) \tag{11.14}$$

The pseudo code of a sequential algorithm implementing the above expression (11.14) is provided below:

```
for (i=1; i<=n; i++) {
    for (j=1; j<=n; j++) {
        c[i,j] = 0;
        for (k=1; k<=n; k++){
            c[i,j]=c[i,j]+a[i,k]*b[k,j];
        }
    }
}
```

Since there is a three-level nested loop in the algorithm, it has a computational complexity of $O(n^3)$. This computational complexity implies that the core operation in the algorithm, `c(i,j)=c(i,j)+a(i,k)*b(k,j)`, has to be executed n^3 times.

An absolute requirement for any parallel algorithm to perform well is that all operands, in this example the matrix elements (a_{ik} and b_{kj}), are available whenever they are needed. This way the data availability will not be the cause of a bottleneck in the computation. The development of a parallel algorithm involves a few issues that do not appear in the sequential algorithm. One of the most significant differences is that we have to decide how the computations and data are distributed to the PEs. Since data movements between PEs are generally expensive, they should be minimized by the initial distribution. We will, in Chapter 12, discuss how the analysis of data dependency can be used to determine the distribution of computations and their data. Currently in this example we simply analyze algorithms that have been developed for parallel architectures.

Two parallel algorithms are presented for this purpose. Note that there are n^2 independent calculations (one for each element in the product matrix) in the matrix multiplication; so potentially n^2 PE can be used.

Algorithm 1

In the first algorithm n PEs (PE$_1$ ... PE$_n$) are available. This algorithm first stores the entire matrix $[A]$ ($a_{11}, ..., a_{ij}, ..., a_{nn}$) in each PE. The kth column of matrix $[B]$ ($b_{1k}, b_{2k}, ..., b_{nk}$) is stored in PE$_k$, $k = 1$ to n. For a reason which will be clear in a moment, the product matrix $[C]$ will also have its kth column ($c_{1k}, c_{2k}, ..., c_{nk}$) stored in PE$_k$, $k = 1$ to n. This initial distribution of matrix elements is shown in Fig. 11.28.

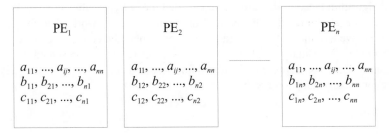

Fig. 11.28 Initial data distribution for matrix multiplication.

The computation then begins as described in the following pseudo code:

```
for (i=1; i<=n; i++){
    concurrently for (k=1; k<=n; k++){
        c[i,k]=0;
    }
    for (j=1; j<=n; j++){
        concurrently for (k=1; k<=n; k++){
            c[i,k]=c[i,k]+a[i,j]*b[j,k];
        }
    }
}
```

In the above pseudo code, the notation of "concurrently for (k=1; k<=n; k++)" describes that all n iterations of the for-loop ($1 \le k \le n$) are carried out concurrently in the n PEs of the system. After the initial distribution of the matrix elements, no additional data transfer is needed during the computation. This algorithm contains a two-level nested sequential loop and thus there are a total of n^2 parallel multiplication. The algorithm has a computational complexity of O(n^2). This complexity does not take into account the initial data distribution and final data collection. The cost of these communications depends on how the PEs are interconnected and also on how they communicate with the host processor. Most likely the parallel architecture serves as a backend processor for the host processor. The host processor is responsible for receiving matrixes [A] and [B] from the user and returning product matrix [C].

The investment of n PEs in this parallel architecture has brought the computational complexity down from $O(n^3)$ of the sequential algorithm to $O(n^2)$. Note that this is not automatically equivalent to a claim that states the parallel system has a speed-up of n. The actual performance gain depends on the overhead of controlling the system and the distribution of data.

Algorithm 2

The performance can be further enhanced if we allow n^2 PEs to be used. In the second algorithm these n^2 PEs are interconnected into a hypercube network. We have discussed the structure of a hypercube interconnection network. For the sake of illustration, we assume that $n = 4$ and present the hypercube connection pattern of 16 PEs in Fig. 11.29. Binary numbers are used to label the PEs represented by circles. Interconnections are represented by double-arrowed links to indicate they are bidirectional communication paths.

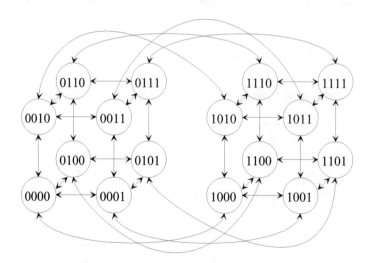

Fig. 11.29 A hypercube interconnection of 16 PEs.

Again, a major part of the algorithm is dedicated to the data distribution. The distribution of the matrix elements is summarized in Fig. 11.30.

Matrix Elements	PE
Row 1 of [A]: $a_{11}, a_{12}, a_{13}, a_{14}$	0000
Row 2 of [A]: $a_{21}, a_{22}, a_{23}, a_{24}$	0101
Row 3 of [A]: $a_{31}, a_{32}, a_{33}, a_{34}$	1010
Row 4 of [A]: $a_{41}, a_{42}, a_{43}, a_{44}$	1111
Column 1 of [B]: $b_{11}, b_{21}, b_{31}, b_{41}$	0000
Column 2 of [B]: $b_{12}, b_{22}, b_{32}, b_{42}$	0100
Column 3 of [B]: $b_{13}, b_{23}, b_{33}, b_{43}$	1000
Column 4 of [B]: $b_{14}, b_{24}, b_{34}, b_{44}$	1100

Fig. 11.30 The distribution of matrix elements in a hypercube architecture.

After the initial data distribution, the rows of [A] are copied from their original PE location to other PEs. Row 1 of [A] is copied from PE_{0000} to PE_{1000}. This takes 4 steps since row 1 of [A] has four elements. Both PE_{0000} and PE_{1000} then copy row 1 of [A] to PE_{0100} and PE_{1100}, respectively. Concurrently with the distribution of row 1, the other three rows of [A] are similarly distributed to four PEs each. The result of these distributions is shown in Fig. 11.31. A similar data distribution is also performed for the columns of [B]. The final distribution of [B] is shown in Fig. 11.32.

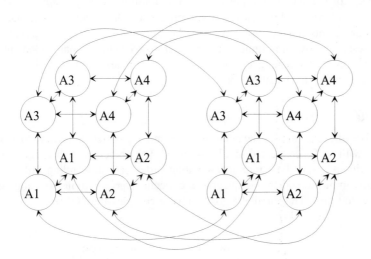

Fig. 11.31 The final distribution of the rows of [A]. A1, A2, A3, and A4 are the first, second, third, and fourth rows of [A], respectively.

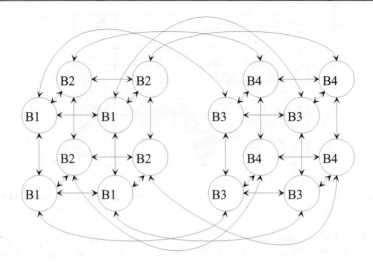

Fig. 11.32 The final distribution of the rows of [B]. B1, B2, B3, and B4 are the first, second, third, and fourth columns of [B], respectively.

Now each PE possesses a row i from [A] and a column j from [B] so that the corresponding element c_{ij} can be calculated. All PEs work in parallel so that the complexity for the actual multiplication and addition is $O(n)$. In addition, the initial distribution of matrix elements has a complexity of $O(n \times \log_2 n)$. The number of steps needed to distribute a row or column is $\log_2 n$ since at each step the number of PEs possessing a row or column doubles. Each row or column has n elements. This algorithm has an interesting characteristic. The complexity of data distribution is higher than that of the actual computation.

Sorting

Before the end of this chapter, we present a parallel algorithm that sorts n numbers with n PEs connected in a linear array. The objective of this algorithm is to sort a series of n numbers into an ascending order or a descending order. Sorting algorithms implemented in a sequential architecture have complexity ranging from $O(n\log_2 n)$ to $O(n^2)$. This parallel algorithm has a complexity of $O(n)$.

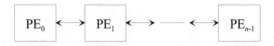

Fig. 11.33 A linear array of n PEs for the parallel sorting algorithm.

The array of n PEs set up for this algorithm is shown in Fig. 11.33. The PEs are connected in a linear array so that each PE (except the PEs on both ends) is connected to two neighbor PEs. The sorting is performed as described in the pseudo code below, assuming that n is an even number. In this algorithm, the n numbers to be sorted are distributed initially into the n PEs, one number per PE. The number currently held by PE$_j$ is represented by v[j].

```
for (i=1; i<n+1; i++){
    if (i is odd){
        concurrently for (j=0; j<n; j++){
            if (j is even){
                if v[j]>v[j+1]{
                    temp=v[j];
                    v[j]=v[j+1];
                    v[j+1]=temp;
                }
            }
        }
    }
    else{
        concurrently for (j=0; j<n-1; j++){
            if (j is odd){
                if v[j]>v[j+1]{
                    temp=v[j];
                    v[j]=v[j+1];
                    v[j+1]=temp;
                }
            }
        }
    }
}
```

Fig. 11.34 provides the snapshots of the array used to sort six numbers distributed initially to the 6 PEs. Six steps are needed to sort them into ascending order as shown in the last snapshot. Note that the longest path of moving a number in a linear array is $(n-1)$ steps. This implies that no algorithm with a complexity lower than $O(n)$ can be found with this parallel architecture.

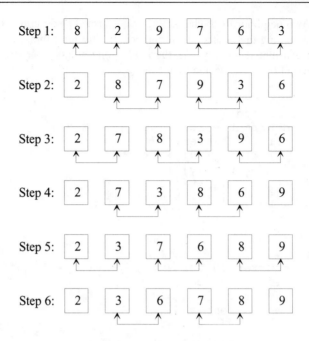

Fig. 11.34 Snapshots of sorting 6 numbers into an ascending order in a linear array.

11.6 Summary

The high density and low cost of VLSI circuits allow innovative algorithms and architectures to be used in problem solving. VLSI processors often explore the parallelism in an application to gain significant speed-ups. Ideally a parallel architecture with n PEs would compute n times faster. Practically this rarely happens and a speed between $\log_2(n)$ and $n/\ln(n)$ is acceptable.

A VLSI architecture designer has to decide the optimal number of PEs and their interconnections. Inter-PE communication is expensive; so a careful analysis has to be done. The data transfer between a host processor and the PEs also has to be well thought of.

General purpose parallel processing existed before VLSI technology dominated the computer industry. In this chapter we presented a few parallel approaches originally developed for parallel computers. These examples provide our further discussion with a reference point. These general purpose parallel computers, in contrast to the VLSI architectures that we will discuss, typically have only a few (e.g., < 10) of relatively powerful PEs (e.g., general central processing units). Our objective is to develop application specific processors that have a larger number of simpler PEs (e.g., at the level of multipliers, adders, etc.).

11.7 To Probe Further

Parallel Processing:

- R. Miller and Q. F. Stout, *Parallel Algorithms for Regular Architectures*, The MIT Press, 1996.

- G. S. Almasi and A. Gottlieb, *Highly Parallel Computing*, The Benjamin/Cummings Publishing Company, 1994.

- M. J. Quinn, *Parallel Computing Theory and Practice*, McGraw-Hill, 1994.

- K. Hwang and F. A. Briggs, *Computer Architectures and Parallel Processing*, McGraw-Hill, 1984.

Interconnection Networks:

- L. N. Bhuyan, Q. Yang, and D. P. Agrawal, "Performance of multiprocessor interconnection networks," *IEEE Computer*, February 1989, pp. 25-37.

11.8 Problems

11.1 Discuss whether or not it is a sound decision to build a parallel architecture to perform the following operation:
$S = a[0] \times b[0] + a[1] \times b[1] + \cdots + a[n] \times b[n]$, where $a[0]$, $a[1]$, ..., $a[n]$, $b[0]$, $b[1]$, ..., $b[n]$ are a series of sequential measurements.

11.2 Repeat Problem 11.1 except that $a[0]$, $a[1]$, ..., $a[n]$, $b[0]$, $b[1]$, ..., $b[n]$ are measurements concurrently made at different locations.

In Problems 11.3–11.7, assume that each addition takes 30 ns, each multiplication takes 50 ns, and the transfer of each datum between two connected PEs takes 10 ns. The times for all other operations (such as memory access, instruction fetch, data fetch, etc.) can be ignored.

11.3 Determine the computational time required to perform the operation given in Problem 11.1 in a system with a single PE. This PE can perform either addition or multiplication but it can only do one operation at a time.

11.4   Repeat Problem 11.3 but consider that the PE has two independent functional units so that addition and multiplication can be carried out at the same time.

11.5   Repeat Problem 11.3 for a system that has n PEs which are connected into a linear array. The interconnecting pattern of a linear array is shown below.

11.6   Repeat Problem 11.5 but consider that the PEs are connected into a ring array.

11.7   Repeat Problem 11.5 but consider that the PEs are fully connected. In other words, each PE is directly connected with any other PE.

11.8   Discuss a few *interesting* applications for an MISD (multiple instruction stream, single data stream) processor.

11.9   Develop an algorithm to determine the routing steps required to send a message from PE_i to PE_j in the interconnection network shown in Fig. 11.9.

11.10   Repeat Problem 11.9 with the interconnection network shown in Fig. 11.10.

11.11   Develop the routing function(s) for an n level binary tree interconnection network.

11.12   Design the multiplexing and demultiplexing structure to be used in a non-linear pipeline such as the one shown in Fig. 11.25.

11.13   An array processor with 16 PEs interconnected into a mesh is shown in Fig. 11.35. Each PE_i is initially distributed with a number v_i. Develop a parallel algorithm to obtain the result

$$\sum_{i=0}^{15} v_i$$

Use "concurrently for" to indicate parallel operations.

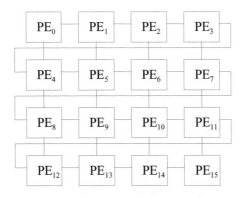

Fig. 11.35 16 PEs connected in a mesh.

Chapter 12   Array Processors

New frontier...

The development in VLSI technology has revolutionized the electronics industry. This is evident in many wonderful devices such as palm-size computers and cellular phones, of which VLSI is the enabling technology.

In order to fully explore the potential of VLSI technology, we are interested in the design and application of massively-parallel, application-specific processors, which are made possible only by the extremely high component density available on a VLSI chip. These processors are built by interconnecting an array of processing elements (PEs) so they are referred to as array processors. This chapter presents the design methodology of high-performance, low-cost, special-purpose array processors.

Comparing array processors to general-purpose parallel processors (e.g., the ones described in Chapter 11) that are mostly based on the concept of datapath and control architecture, array processors have the following properties:

- Simple and small PEs allow their designs to be optimized. A full-custom PE design process is often justifiable.
- Only a few different types of PEs are required. This property reduces the design cost.
- Simple and regular interconnections between PEs simplify the placement and routing process. Array processors often have mostly nearest neighbor interconnections.
- Extensive parallel processing and pipelining are employed to sustain a high throughput.
- The data flow often follows a simple and regular pattern.
- Very few explicit global control signals are needed.
- The clock signal and power are distributed in a regular pattern. This property allows the optimization of clock and power trees.

Systolic arrays and wavefront arrays will be discussed in this chapter. Systolic arrays are synchronous and wavefront arrays are asynchronous. In a synchronous systolic array, the operations of all PEs are synchronized using a common clock signal. The distribution of a clock signal in a large array must be designed cautiously to avoid clock skews.

In contrast, asynchronous wavefront arrays use local clocks. The coordination between PEs is done by handshaking. The operation of a PE in an asynchronous array is activated when all its inputs are available.

In systolic array processors, data must be fed into the system according to a carefully planned schedule. Asynchronous wavefront array processors, on the other hand, are self-timed so scheduling is relatively simple. In either case, data are processed at some PEs before they are transferred to other PEs for further processing. In this manner, data are continuously pumped through an array processor to generate a very high throughput rate.

Before we explain the design process of array processors, a few examples of array processing will be discussed.

12.1 Array Processing Examples

Matrix-Vector Multiplication

Vector and matrix operations are the cornerstone building blocks of many digital signal processing (DSP) algorithms. For example, digital filtering, which has many applications in DSP, is based on inner product computation.

In this example we discuss array processing structures that compute the product $Y = [y_1, y_2, y_3, y_4]^T$ of a matrix $A = [a_{ij}]_{4 \times 4}$ and a vector $X = [x_1, x_2, x_3, x_4]^T$:

$$
\begin{bmatrix} y_1 \\ y_2 \\ y_3 \\ y_4 \end{bmatrix} = \begin{bmatrix} a_{11} & a_{12} & a_{13} & a_{14} \\ a_{21} & a_{22} & a_{23} & a_{24} \\ a_{31} & a_{32} & a_{33} & a_{34} \\ a_{41} & a_{42} & a_{43} & a_{44} \end{bmatrix} \times \begin{bmatrix} x_1 \\ x_2 \\ x_3 \\ x_4 \end{bmatrix}
\tag{12.1}
$$

The algorithm for computing a component $y_i = \sum_{k=1}^{4} a_{ik} x_k$ in Y is described in the following pseudo code:

```
y[i] = 0;
for (k=1; k<n+1; k++){
    y[i] = y[i] + a[i][k]*x[k];
}
```

The computation described in the above program can be implemented as a pipeline operation, in which y_i (y[i]) is initiated to be 0 and is streamlined through a pipeline. In its journey y_i collects all $a_{ik} x_k$ (a[i][k]*x[k]) terms and evolves into the final result. This observation suggests the multiple pipeline structure shown in Fig. 12.1.

Four pipelines, one for each y_i, are provided to compute Y. The function of a stage is also shown in Fig. 12.1. Each stage of a pipeline stores the appropriate a_{ik} and x_i initially distributed to it for the computation of $y_i' = y_i + a_{ik} x_k$. If a stage computes y_i' in 1 clock cycle (clock not shown), the entire operation takes 4 clock cycles to complete. The throughput of this multiple pipeline structure is 4 times over that of a single processor. However, the hardware cost is 16 times over that of a single processor.

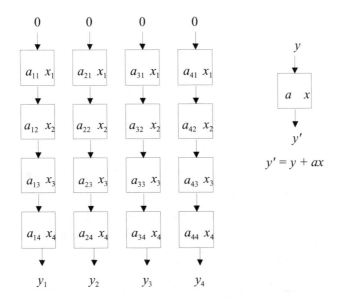

Fig. 12.1 Multiple pipeline structure for matrix-vector multiplication.

A further consideration will reveal that the flows of x_k's and a_{ik}'s can also be streamlined. We modify the pipeline architecture into the structure shown in Fig. 12.2. We will discuss in detail how this structure is designed in a moment. At this time our objective is to appreciate its operation.

The function of each PE is also shown in Fig. 12.2. As in a pipeline structure, each interconnection between PEs is provided with a register which regulates the flow of data. For the purpose of illustration, we show the connection between y' and y as a feedback. In practice, this feedback loop can be implemented as an internal register of the PE and is updated at each clock cycle. The operations of this array processor are demonstrated in Fig. 12.3 to Fig. 12.7.

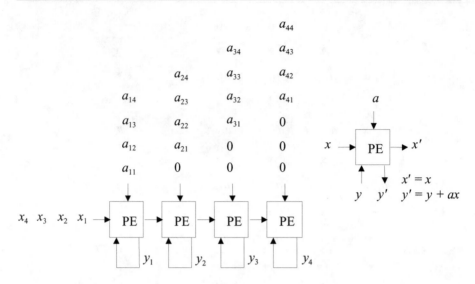

Fig. 12.2 Array processor for matrix-vector multiplication.

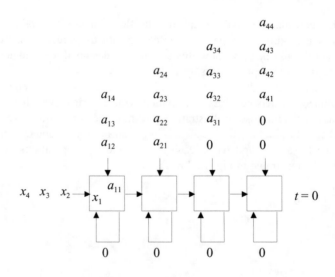

Fig. 12.3 Snapshot of the array processor at $t = 0$.

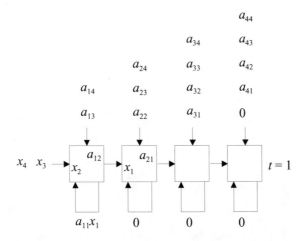

Fig. 12.4 Snapshot of the array processor at $t = 1$.

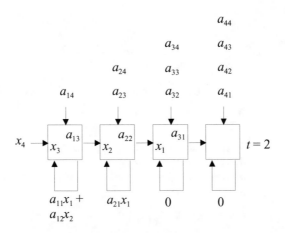

Fig. 12.5 Snapshot of the array processor at $t = 2$.

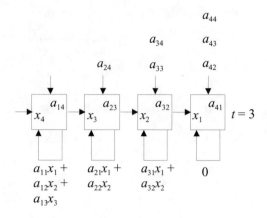

Fig. 12.6 Snapshot of the array processor at $t = 3$.

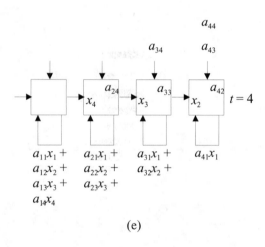

(e)

Fig. 12.7 Snapshot of the array processor at $t = 4$.

Fig. 12.3 to Fig. 12.7 shows the snapshots of the array processor in the first 4 clock cycles of its operation. Note that the intermediate results are stored in the feedback loops (or internal registers). The first Y element y_i is available after 4 clock cycles. From this point on, it takes one additional cycle for the next element to be produced. The entire result takes 7 clock cycles to complete.

The following example shows a variation of the above matrix-vector multiplication. The new problem is to multiply a band matrix $A = [a_{ij}]$ with a vector $X = [x_1, x_2, \ldots]^T$. This problem also has many applications (e.g., finite impulse response filtering) in digital signal processing. The operation to be performed is mathematically described in (12.2).

$$
\begin{bmatrix} y_1 \\ y_2 \\ y_3 \\ y_4 \\ \cdot \\ \cdot \end{bmatrix} = \begin{bmatrix} a_{11} & a_{12} & & & & & 0 \\ a_{21} & a_{22} & a_{23} & & & \\ a_{31} & a_{32} & a_{33} & a_{34} & & \\ & a_{42} & a_{43} & a_{44} & a_{45} & \\ & & a_{53} & & \cdot\cdot & \\ 0 & & & & & \cdot\cdot \end{bmatrix} \times \begin{bmatrix} x_1 \\ x_2 \\ x_3 \\ x_4 \\ \cdot \\ \cdot \end{bmatrix} \tag{12.2}
$$

The main difference between this computation and the previous one is that the dimension of the matrix is not fixed. Vector X can be considered as a stream of input data. Similarly, vector Y is a stream of outputs. Each row in the matrix A represents the coefficients of a filtering operation. A few of the results are:

$$
\begin{aligned}
y_1 &= a_{11}x_1 + a_{12}x_2; \\
y_2 &= a_{21}x_1 + a_{22}x_2 + a_{23}x_3; \\
y_3 &= a_{31}x_1 + a_{32}x_2 + a_{33}x_3 + a_{34}x_4; \\
y_4 &= a_{42}x_2 + a_{43}x_3 + a_{44}x_4 + a_{45}x_5;
\end{aligned}
$$

...

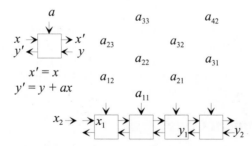

Fig. 12.8 Linear array processor for matrix-vector multiplication.

A linear array processor designed to carry out this operation is shown in Fig. 12.8. Note the directions of data movement in this array processor, in which x_k moves from left to right, y_i moves from right to left, and a_{ik}'s move from top to bottom. The scheduling of these data has been carefully planned so that right data will meet in the right place at the right time. A sequence of snapshots of this array processor are shown in Fig. 12.9.

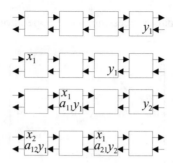

Fig. 12.9 Snapshots of the array processor shown in Fig. 12.8.

The reader should follow the data flow of this array processor to develop a mental picture of its operation. Several observations can be made to this array processor. The number of PEs in the processor equals the bandwidth of the matrix (i.e., $n = 4$ in this case). Similar to the operation of a pipeline, the first result will take n cycles to appear at the left end of the array. The other results are produced at the rate of 1 result every 2 clock cycles.

Since the array processor can only take 1 input data every 2 clock cycles, it has to be operated at a clock frequency twice that of the input data rate. For example, if the data rate is 100 million samples per second (MSPS), then the array processor should be operated at 200 MHz.

12.2 Array Processor Design

A VLSI architecture can be developed by exploring the parallelism in a given algorithm. Once the parallelism is detected, the algorithm can be converted into the description of a physical PE array. The number of PEs in the array, the capability of each PE, their interconnection pattern, and the data flow can be determined. A VLSI chip is created to perform the computation.

Single Assignment Code

An algorithm is normally written as a sequence of steps. With the exception of a few programming languages developed specifically for parallel processing, most languages do not have the capability to express parallelism. The consequence of this limitation is that unnecessary ordering of operations is imposed on the computation. For example, consider the addition of two matrices

$$\begin{bmatrix} c_{11} & c_{12} & \cdots & c_{1n} \\ c_{21} & c_{22} & \cdots & c_{2n} \\ \vdots & \ddots & & \vdots \\ c_{n1} & c_{n2} & \cdots & c_{nn} \end{bmatrix} = \begin{bmatrix} a_{11} & a_{12} & \cdots & a_{1n} \\ a_{21} & a_{22} & \cdots & a_{2n} \\ \vdots & \ddots & & \vdots \\ a_{n1} & a_{n2} & \cdots & a_{nn} \end{bmatrix} + \begin{bmatrix} b_{11} & b_{12} & \cdots & b_{1n} \\ b_{21} & b_{22} & \cdots & b_{2n} \\ \vdots & \ddots & & \vdots \\ b_{n1} & b_{n2} & \cdots & b_{nn} \end{bmatrix} \quad (12.3)$$

When coded in a sequential programming language, the operation in (12.3) becomes

```
for (i=1, i≤n, i++){
    for (j=1, j≤n, j++){
        c[i][j]=a[i][j]+b[i][j];
    }
}
```

In the above pseudo code, the elements of matrices A and B are accessed in a row major order. This decision is completely arbitrary. Indeed, there is no gain or loss if we swap the two "for" statements so that the pseudo code reads

```
for (j=1, j≤n, j++){
    for (i=1, i≤n, i++){
        c[i][j]=a[i][j]+b[i][j];
    }
}
```

After the swapping, the elements of matrices A and B are now accessed in a column major order. This type of implied ordering is unnecessarily presented in these sequential codes. The first step in designing an array processor is to remove unneeded ordering from a given algorithm.

Consider a generalized matrix-vector multiplication problem as described below:

$$\begin{bmatrix} y_1 \\ y_2 \\ \vdots \\ y_n \end{bmatrix} = \begin{bmatrix} a_{11} & a_{12} & \cdots & a_{1m} \\ a_{21} & a_{22} & \cdots & a_{2m} \\ \vdots & \vdots & \ddots & \vdots \\ a_{n1} & a_{n2} & \cdots & a_{nm} \end{bmatrix} \times \begin{bmatrix} x_1 \\ x_2 \\ \vdots \\ x_m \end{bmatrix} \quad (12.4)$$

The computation of the product vector Y can be expressed in the following pseudo code:

```
for (i=1; i<=n; i++){
    y[i] = 0;
    for (j=1; j<=m; j++){
        y[i] = y[i]+a[i][j]*x[j];
    }
}
```

}

According to the ordering implied in the above pseudo code, y [2] (i.e., y_2) is computed after y [1]. This ordering is absolutely unnecessary since there is no data dependence between y [2] and y [1]. In other words, the computation of y [2] does not depend on the result of computing y [1], and vice versa. Similarly, it is not necessary to compute a [i] [1] × b [1] before a [i] [2] × b [2], and vice versa. This kind of artificial dependence must be removed in the description of an algorithm for VLSI implementation, in which parallel processing is to be emphasized.

We can remove the unnecessary dependence in the above pseudo code by introducing a syntax that allows parallel operations to be explicitly specified. The result is shown in the following pseudo code:

```
parallel for (i=1; i<=n; i++) and (j=1; j<=m; j++){
    temp[i][j] = a[i][j] * x[j];
}
parallel for (i=1; i<=n; i++){
    y[i]=0;
    for (j=1; j<=m; j++){
        y[i] = y[i] + temp[i][j];
    }
}
```

The first three lines in the pseudo code describe that all $n \times m$ multiplications (a [i] [j] *b [j]) can be carried out independently and thus should be computed in parallel. Similarly, the formation of the components in the result vector c can also be performed in parallel. This is specified in the rest of the pseudo code. Our conclusion from this example is that unnecessary ordering of operations can be eliminated by studying the data dependence of the operations.

Another inadequacy of a sequential programming language is that it uses the assignment operator (=) to indicate ordering. For example, in the line y [i] = y [i]+temp [i] [j], the term y [i] on the left side (receiver side) of the equal sign is not available until the operation on the right side (feeder side) of the equal sign is complete. The same symbol (y [i]) is used to represent two different values. In a sequential programming, this is not a problem as long as we observe the convention of using the equal sign.

In a hardware implementation, using a symbol to represent different values implies the use of a variable, possibly stored in a register. In order to clearly distinguish these values in the exploration of parallelism, we can introduce a time dimension into the description and make it recursive. For example, we can rewrite the statement as

y [i] [k+1] = y [i] [k] + temp [i] [j];

and use a loop involving index k to describe the operations. The entire pseudo code is now rewritten as:

```
for (i=1; i<=n; i++){
    y[i][1] = 0;
    for (j=1; j<=m; j++){
        y[i][j+1] = y[i][j] + a[i][j] * x[j];
    }
}
```

Note that we have taken advantage of the fact that the newly introduced index k always equal to j in the computation. This allows index j doubled as the "time" index. The latest version of the matrix-vector multiplication pseudo code is an example of single-assignment code. In single-assignment code, no variable is assigned with values more than once. The creation of a single-assignment code for a given computation is the first step of developing an array processor for it. Based on single assignment code, the dependence between data is then explored for maximum parallelism.

Dependence Graph

The generation of a dependence graph is the next step after the code describing an algorithm is described as a single assignment code. Dependence graph is a tool for exploring the parallelism in an algorithm. Operations of an algorithm that have no dependence on each other can be executed concurrently in a parallel architecture.

A dependence graph is a graph that shows the dependence of computations occurring in an algorithm. The dependence graph of a single assignment code does not have loops or cycles.

The data dependence graph for the operation y[i][j+1] = y[i][j] + a[i][j] * b[j] is shown in Fig. 12.10. The indices (i, j) are used to describe a two-dimensional space so that each pair of i and j become the coordinates in this space. We assume $n = m = 4$ in this example. The arrows in the dependence graph are simply the dependence described in the single assignment statement. Notice the similarity between Fig. 12.1 and Fig. 12.10. In fact, each circle in Fig. 12.10 is equivalent to a stage in the multiple pipeline structure in Fig. 12.1. Now we see that the multiple pipeline structure is a direct implementation of the dependence graph of matrix-vector multiplication.

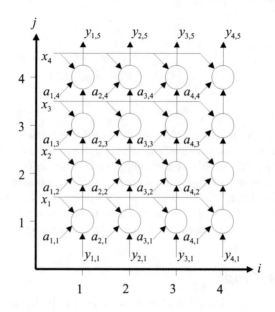

Fig. 12.10 Dependence graph for matrix-vector multiplication.

The interconnections passing y's are local nearest neighbor interconnections. The x's, on the other hand, are global interconnections, which may be costly to implement in a VLSI architecture. We thus modify the dependence graph of Fig. 12.10 to convert the global interconnections into local ones. The result is shown in Fig. 12.11.

The following pseudo code would have been the algorithm described by the modified dependence graph in Fig. 12.11.

```
for (i=1; i<=4; i++){
    y[i][1] = 0;
    for (j=1; j<=4; j++){
        x[i+1][j] = x[i][j];
        y[i][j+1] = y[i][j] + a[i][j] * x[i][j];
    }
}
```

The statement `x[i+1][j]=x[i][j]` converts the global interconnections into local ones by introducing index i to represent a time dimension.

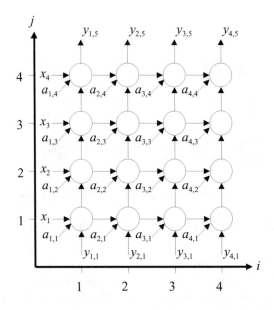

Fig. 12.11 Modified dependence graph with local interconnections.

Signal Flow Graph

While a dependence graph suggests a structure to implement an algorithm, the PEs do not have a high utilization rate. We can improve the PE utilization rate by projecting a data dependence graph into a signal flow graph, which is a more concise representation of the algorithm to be implemented. The result signal flow graph suggests the structure of an array processor. Typically, an n-dimensional dependence graph is projected into a signal flow graph of $(n-1)$ dimensions. An example of projecting a 2-dimensional dependence graph into a 1-dimensional array is shown in Fig. 12.12. Each node in the signal flow graph represents a PE of the array processor.

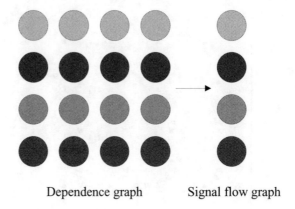

Dependence graph   Signal flow graph

Fig. 12.12 Projection of a dependence graph to a signal flow graph.

Once the structure of the signal flow graph and thus an array processor is determined by the projection, the designer has to answer two questions:
• Which PE in the array processor should an operation (i.e., a node in a dependence graph) be mapped to?
• What is the correct ordering of operations (i.e., the arcs in a dependence graph) in the PEs of the array processor?

In a dependence graph, each node represents a specific operation step. Since we typically have less nodes in the array processor than in the data dependence graph, the operations in more than one node have to be mapped to the same PE. The scheduling of these operations mapped to the same PE will have to be determined.

When there is more than one way of projecting a dependence graph into an array processor, we can evaluate the quality of the projection by considering:
• the throughput, which is defined as the number of results produced per unit area;
• the PE utilization rate, which is defined as the throughput divided by the number of PEs; and
• the complexity of PE interconnection.

Before we get in the details of projecting a data dependence graph into a signal flow graph, we define a few vectors in the data dependence graph space to facilitate the description of projection.

Fig. 12.13 Example vectors for direction representations.

Four direction representing vectors (1–4) are shown in Fig. 12.13. Arrow 1 points in the j direction and is represented by the vector

$$\begin{bmatrix} i \\ j \end{bmatrix} = \begin{bmatrix} 0 \\ 1 \end{bmatrix}$$

The other three directions (arrows 2, 3, and 4) are represented by

$$\begin{bmatrix} i \\ j \end{bmatrix} = \begin{bmatrix} 1 \\ 1 \end{bmatrix}, \begin{bmatrix} -1 \\ -1 \end{bmatrix}, \text{ and } \begin{bmatrix} 1 \\ 0 \end{bmatrix}$$

respectively. The use of these direction vectors will be demonstrated in a moment.

In the projection of a dependence graph into an array processor structure, a projection vector P has to be determined. The second step is to select a permissible schedule for the operations represented by the nodes of a data dependence graph. This can be done by analyzing the dependence graph to identify a set of parallel and equally spaced hyperplanes. The dependence graph nodes on the same hyperplane identify operations that can be executed in parallel in the result array processor. Different hyperplanes represent different clock cycles of operations.

In order for a schedule to be permissible, the relationship of data dependence must not be violated. For example, if node m depends on node n, then node m cannot be located on a hyperplane corresponding to a time step earlier than the hyperplane of node n. Also, the hyperplanes must not be parallel with the projection vector P. This avoids the mapping of concurrent operations to the same PE. Parallelism can then be maximized.

As an example, a projection vector

$$P = \begin{bmatrix} 0 \\ 1 \end{bmatrix}$$

for the matrix-vector multiplication dependence graph (Fig. 12.10) is compatible with the hyperplane set shown in Fig. 12.14. The hyperplanes are shown in dashed lines. We refer to a hyperplane set by specifying a vector S that points in the direction of time flow.

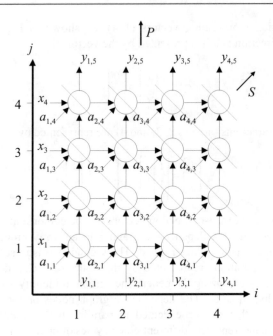

Fig. 12.14 Valid projection vector and hyperplane set.

In Fig. 12.14,

$$S = \begin{bmatrix} 1 \\ 1 \end{bmatrix}$$

which indicates that the flow of data is going from the lower left corner of the graph toward the upper right corner. Note that this is not the only valid hyperplane set for this data dependence graph. In a 2-dimensional data dependence graph, there are eight possible vectors. In addition to S, another valid hyperplane set is

$$S_2 = \begin{bmatrix} 0 \\ 1 \end{bmatrix}$$

Invalid hyperplanes are

$$S_3 = \begin{bmatrix} 1 \\ 0 \end{bmatrix} \text{ (hyperplanes are in parallel with } P\text{),}$$

$$S_4 = \begin{bmatrix} -1 \\ -1 \end{bmatrix} \text{ (data dependence violated)},$$

$$S_5 = \begin{bmatrix} 0 \\ -1 \end{bmatrix} \text{ (data dependence violated)},$$

$$S_6 = \begin{bmatrix} -1 \\ 0 \end{bmatrix} \text{ (data dependence violated)},$$

$$S_7 = \begin{bmatrix} -1 \\ 1 \end{bmatrix} \text{ (data dependence violated)},$$

$$S_8 = \begin{bmatrix} 1 \\ -1 \end{bmatrix} \text{ (data dependence violated)}.$$

In summary, the selection of a projection vector and a set of hyperplanes must satisfy the following conditions:
1. Data dependence must not be violated.
2. The projection vector and the hyperplanes must not be in parallel.

These two conditions can be expressed mathematically as follows. The first condition can be described as

$$S^T \times e \geq 0 \qquad\qquad (12.5)$$

in which S^T is the transpose of vector S, and e is the direction of any dependence arc in the dependence graph. Satisfying (12.5) implies that no dependence arcs are going against the flowing direction of the hyperplanes.

For example, in the dependence graph of Fig. 12.14, there are two types of dependence arcs:

$$e_1 = \begin{bmatrix} 1 \\ 0 \end{bmatrix} \text{ and } e_2 = \begin{bmatrix} 0 \\ 1 \end{bmatrix}$$

Checking the validity of S, we have

$$S^T \times e_1 = \begin{bmatrix} 1 & 1 \end{bmatrix} \times \begin{bmatrix} 1 \\ 0 \end{bmatrix} = 1 \geq 0$$

$$S^T \times e_2 = \begin{bmatrix} 1 & 1 \end{bmatrix} \times \begin{bmatrix} 0 \\ 1 \end{bmatrix} = 1 \geq 0$$

This proves the hyperplanes represented by S satisfy (12.5). On the other hand, checking the validity of

$$S_5 = \begin{bmatrix} 0 \\ -1 \end{bmatrix}$$

we found that

$$S_5^T \times e_1 = \begin{bmatrix} 0 & -1 \end{bmatrix} \times \begin{bmatrix} 1 \\ 0 \end{bmatrix} = 0$$

$$S_5^T \times e_2 = \begin{bmatrix} 0 & -1 \end{bmatrix} \times \begin{bmatrix} 0 \\ 1 \end{bmatrix} = -1$$

Hyperplane set S_5 hence does not satisfy (12.5).

The second condition, which implies that the hyperplanes are not in parallel with the projection direction, can be described as

$$S^T \times P > 0 \qquad\qquad (12.6)$$

Checking the relationship between S and P in Fig. 12.14, we found that

$$S^T \times P = \begin{bmatrix} 1 & 1 \end{bmatrix} \times \begin{bmatrix} 0 \\ 1 \end{bmatrix} = 2 > 0$$

so (12.6) is satisfied. If we apply the test to

$$S_3 = \begin{bmatrix} 1 \\ 0 \end{bmatrix}$$

which is in parallel to P, we have

$$S_3^T \times P = \begin{bmatrix} 1 & 0 \end{bmatrix} \times \begin{bmatrix} 0 \\ 1 \end{bmatrix} = 0$$

which violates the condition of (12.6).

Node Mapping

Given a projection vector P, it is a straightforward exercise to map the node activities in a data dependence graph into an array processor. We continue to use the matrix-vector multiplication example to illustrate this mapping.

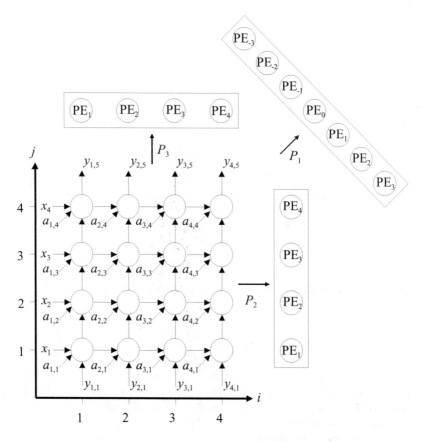

Fig. 12.15 Three different projections of a dependence graph.

Fig. 12.15 shows the use of three different projection vectors (P_1, P_2, and P_3) to assign node activities into PEs. Each of these projections is valid. P_1 projects the data dependence graph into a linear array of 7 PEs. P_2 and P_3 project the data dependence into linear arrays of 4 PEs. The interconnections between the PEs in an array are to be determined in a moment.

The mapping of node activities according to a selected projection vector P can be determined by

$$c' = H^T \times c \qquad (12.7)$$

where c is the location of a node in the data dependence graph, c' is the location of a PE to which the node is to be mapped, H^T is an $n \times (n-1)$ node mapping matrix (n is the number of dimensions in the data dependence graph) that is orthogonal to projection vector P (i.e., $H^T \times P = 0$).

In Fig. 12.15,

$$P_1 = \begin{bmatrix} 1 \\ 1 \end{bmatrix}$$

We select

$$H_1 = \begin{bmatrix} 1 \\ -1 \end{bmatrix}$$

so that $H_1^T \times P_1 = 0$. Each node in the data dependence graph can be identified as

$$\begin{bmatrix} i \\ j \end{bmatrix}$$

The mapping of the nodes into the array processor created according to P_1 is found to be

$$c' = \begin{bmatrix} 1 & -1 \end{bmatrix} \times \begin{bmatrix} i \\ j \end{bmatrix} = i - j \qquad (12.8)$$

For example, all nodes with $i = j$ (i.e., (1, 1), (2, 2), (3, 3), and (4, 4)) are mapped into PE_0. Node (2, 1) is mapped into PE_1. Note that the PEs are identified by their positional relationship to PE_0.

If a different projection direction

$$P_2 = \begin{bmatrix} 1 \\ 0 \end{bmatrix}$$

is chosen, we select

$$H_2 = \begin{bmatrix} 0 \\ 1 \end{bmatrix}$$

so that $H_2^T \times P_2 = 0$. The mapping of the nodes into the array processor according to this projection vector is found to be

$$c' = \begin{bmatrix} 0 & 1 \end{bmatrix} \times \begin{bmatrix} i \\ j \end{bmatrix} = j \tag{12.9}$$

which means that all activities in a row (i.e., identical j) are mapped to the same j^{th} PE in the array. The mapping carried out with P_3 can be done similarly.

Arc Mapping

The arcs in a data dependence graph describe the dependencies between operations. They are used to determine the interconnection pattern between the PEs in an array processor. While the arc mapping can be carried out manually for a simple data dependence graph, a systematic approach is desirable for more complex operations to avoid the possibility of making errors. In addition to the interconnections between PEs, this step also determines the delays (i.e., registers) that must be provided on an interconnection. The need of delays on interconnections is obvious if we consider the possibility of having a dependence between PEs that operate in parallel. A pipeline structure is an example of such a situation.

In summary, we need to determine two things in this mapping step:

1. The interconnection e' in the array processor corresponding to a dependence e.
2. The number of delays, $D(e')$, that is required in the interconnection e'.

Both of these can be found by the following simple matrix operation:

$$\begin{bmatrix} D(e') \\ e' \end{bmatrix} = \begin{bmatrix} S^T \\ H^T \end{bmatrix} \times e \tag{12.10}$$

where S and H are the normal vector of the hyperplanes and the node mapping matrix, respectively.

Continue with the matrix-vector multiplication example. We have selected a projection vector of $P_1 = [1, 1]^T$. In the node mapping we have determined H to be $[1 \ -1]^T$. We have also chose a set of hyperplanes represented by $S = [1 \ 1]^T$. The mapping of $e_1 = [1 \ 0]^T$ is found as

$$\begin{bmatrix} D(e_1{}') \\ e_1{}' \end{bmatrix} = \begin{bmatrix} 1 & 1 \\ 1 & -1 \end{bmatrix} \times \begin{bmatrix} 1 \\ 0 \end{bmatrix} = \begin{bmatrix} 1 \\ 1 \end{bmatrix} \qquad (12.11)$$

This result shows that each e_1 in the data dependence graph is mapped into an interconnection ($e_1{}'$) between PEs that goes in the position direction (i.e., the interconnection goes from PE_n to PE_{n+1}). A delay ($D(e_1{}')$), which can be implemented with a register, should be inserted into these interconnections.

On the other hand, the mapping of $e_2 = [0\ 1]^T$ is found to be

$$\begin{bmatrix} D(e_2{}') \\ e_2{}' \end{bmatrix} = \begin{bmatrix} 1 & 1 \\ 1 & -1 \end{bmatrix} \times \begin{bmatrix} 0 \\ 1 \end{bmatrix} = \begin{bmatrix} 1 \\ -1 \end{bmatrix} \qquad (12.12)$$

This result shows that each e_2 in the data dependence graph is mapped into an interconnection ($e_2{}'$) between PEs that goes in the negative direction (i.e., the interconnection should goes from PE_n to PE_{n-1}). A delay ($D(e_2{}')$) should be inserted into these interconnections. The result of this mapping is shown in Fig. 12.16.

Fig. 12.16 Mapping of interconnections.

As another example, if P_2 is selected to project the node activities, we have $H = [0\ 1]^T$ and $S = [1\ 1]^T$. The mapping of e_1 is found to be

$$\begin{bmatrix} D(e_1{}') \\ e_1{}' \end{bmatrix} = \begin{bmatrix} 1 & 1 \\ 0 & 1 \end{bmatrix} \times \begin{bmatrix} 1 \\ 0 \end{bmatrix} = \begin{bmatrix} 1 \\ 0 \end{bmatrix} \qquad (12.13)$$

The result of direction '0' indicates that it is going to stay in the same PE to form a feedback loop. One delay should be provided for this feedback loop. The mapping of e_2 is found to be

$$\begin{bmatrix} D(e_2') \\ e_2' \end{bmatrix} = \begin{bmatrix} 1 & 1 \\ 1 & -1 \end{bmatrix} \times \begin{bmatrix} 1 \\ 0 \end{bmatrix} = \begin{bmatrix} 1 \\ 1 \end{bmatrix} \tag{12.14}$$

which indicates an interconnection that goes in the positive direction with one delay. The result is shown in Fig. 12.17.

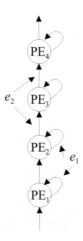

Fig. 12.17 Another interconnection mapping.

I/O Mapping

The last step is to determine when and where to apply inputs and to collect outputs. The PE position c' and the time $t(c')$ to perform I/O operations can be determined by

$$\begin{bmatrix} t(c') \\ c' \end{bmatrix} = \begin{bmatrix} S^T \\ H^T \end{bmatrix} \times c \tag{12.15}$$

where c is the node location in the dependence graph, c' is the PE location, $t(c')$ is the time to apply or collect data, S is the normal vector of the hyperplanes, and H is the node mapping matrix. Note that the $t(c')$ is expressed in a relative time unit.

Continue with the matrix-vector multiplication example. If P_1 is selected, we have the following result. The value $a_{i,j}$ is applied at node $c_1 = [i, j]^T$, thus

$$\begin{bmatrix} t(c_1') \\ c_1' \end{bmatrix} = \begin{bmatrix} 1 & 1 \\ 1 & -1 \end{bmatrix} \times \begin{bmatrix} i \\ j \end{bmatrix} = \begin{bmatrix} i+j \\ i-j \end{bmatrix} \tag{12.16}$$

The value y_i is applied at node $c_2 = [i, 1]^T$, thus

$$\begin{bmatrix} t(c_2') \\ c_2' \end{bmatrix} = \begin{bmatrix} 1 & 1 \\ 1 & -1 \end{bmatrix} \times \begin{bmatrix} i \\ 1 \end{bmatrix} = \begin{bmatrix} i+1 \\ i-1 \end{bmatrix} \qquad (12.17)$$

The values x_i are applied at node $c_3 = [1, j]^T$, thus

$$\begin{bmatrix} t(c_3') \\ c_3' \end{bmatrix} = \begin{bmatrix} 1 & 1 \\ 1 & -1 \end{bmatrix} \times \begin{bmatrix} 1 \\ j \end{bmatrix} = \begin{bmatrix} 1+j \\ 1-j \end{bmatrix} \qquad (12.18)$$

The output y are collected at node $c_4 = [i, 5]^T$, thus

$$\begin{bmatrix} t(c_4') \\ c_4' \end{bmatrix} = \begin{bmatrix} 1 & 1 \\ 1 & -1 \end{bmatrix} \times \begin{bmatrix} i \\ 5 \end{bmatrix} = \begin{bmatrix} i+5 \\ i-5 \end{bmatrix} \qquad (12.19)$$

Fig. 12.18 summarizes these mapping results. The inputs $a_{1,1}$, x_1, and $y_{1,1}$ are applied to PE_0 at $t = 2$. The other inputs are shown in positions that indicate their timing relationship with respect to each other. The outputs are collected at the upper four PEs (PE_{-3} to PE_0) with their available time indicated.

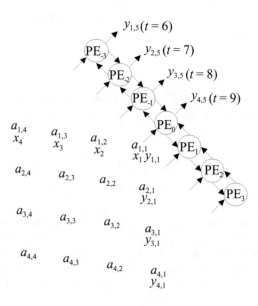

Fig. 12.18 I/O of the array processor.

The completion of the other possible projection directions is left as exercises (see Problem 12.1).

12.3 Wavefront Array Processor

Synchronous array processors offer many benefits such as easy to analyze and at least theoretically easy to control. However, one big drawback is the need of a global clock that has to be distributed through the circuit. Clock-skews due to different clock path lengths and loads have to be carefully avoided.

The use of registers for the purpose of synchronization may cause unnecessary power dissipation. All PEs are switching states at more or less the same time since they are triggered by the same clock signal. This causes large current surges to occur. Sufficient bypass capacitors should be provided on chip to avoid hazards caused by current surges.[1]

The wavefront array processor is a potential solution to the problems that come along with synchronous array processors. Wavefront processor uses the concept of a data-flow machine, which operates only after all operands have arrived. We say that the PE is triggered by the arrivals of operands. The result of an operation, when available, is forwarded to another PE to activate its operation. Global control and synchronization are no longer needed.

Unlike a general purpose data-flow computer which needs a powerful control mechanism to manage data and resources, VLSI array processors have inherent modularity and local communication which allow them to be potential data-flow architectures.

A wavefront array is also called a self-timed, data-driven system. Each PE has I/O buffers and handshaking capability. The arrival of data from a neighbor PE is interpreted as a signal to change state and to trigger new activities. Each PE may contain its own clock. In fact, PEs can operate at different clock rates.

Fig. 12.19 shows the general structure of a PE used in a wavefront array. The PE has two inputs (a and b) and two outputs (c and d). Each I/O is provided with a queue which is indicated by the short cut-lines on the signal line. The size of the queue is represented by the number of cut-lines. For example, input a has a single buffer queue so it can hold at most one datum. The queue of input b has 2 buffers. Similarly, outputs c and d have double-buffered and single-buffered queues, respectively. The data stored in a queue are represented by the shaded circles, which will be referred to as data tokens.

Fig. 12.19(a) shows inputs a and b, each of which has one token in its queue. Output c has one token and one space. A PE is activated if the following two conditions are met:

- Each input has at least one data token.
- Each output has at least one empty space.

---

[1] A rule of thumb for determining the bypass capacitance is that it should be about 10 times of the total capacitance involved with the switching.

It is easy to verify that the PE in Fig. 12.19(a) is in its activated state. The PE in Fig. 12.19(b) is in a waiting state because its input a queue is empty and its output d queue is full.

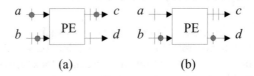

(a)   (b)

Fig. 12.19 PE used in a wavefront array: (a) a PE in activated state; and (b) a PE in waiting state.

Fig. 12.20 shows the operation of a wavefront array PE with a propagation delay of 4 ($\tau = 4$). At $t = 0$, the PE is activated since the activation conditions have been satisfied. It takes 4 time units for the PE to produce results. The state of the PE at $t = 4$ is also shown in Fig. 12.20, in which the tokens at its inputs (a, b) have been consumed and removed from their queues. Results are inserted in the queues of outputs c and d.

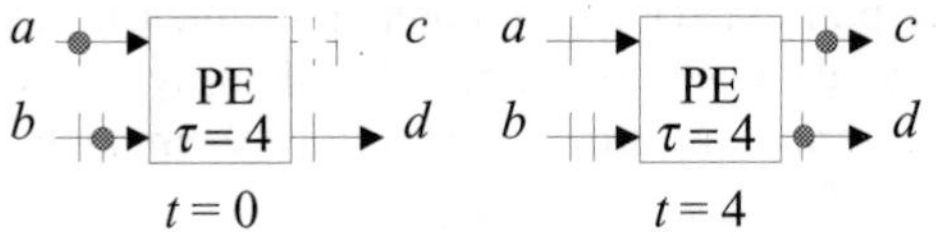

Fig. 12.20 Operations of a wavefront array PE.

Circuitry must be provided to detect the arrival and departure of data tokens. The cost of a wavefront array is mainly in its large overhead. The following rule of thumb can be used to guide the selection between a systolic array and a wavefront array to implement an algorithm. A systolic array is preferred if global clock distribution does not present a problem and the PEs are small, simple modules. A wavefront array is more appropriate in a situation when the PEs are relatively complicated (so that the overhead of handshaking can be ignored) and global synchronization is a problem.

The wavefront implementation of an application can be created in essentially the same way a systolic array is developed. In fact, a wavefront array can be derived from a systolic array intuitively. Instead of using a global clock to synchronize all PEs, the operation is controlled by the movements of data tokens. Note that the scheduling step in a systolic array design is no longer needed. No exact timing specification is needed. The data input to boundary nodes are used as initial data tokens to start the array processor operation.

Two issues have to be addressed in the design of a wavefront array processor. The first is the possibility of a deadlock situation in which one or more PE is stuck in a waiting state indefinitely. Fig. 12.21 shows a wavefront array processor that is in a deadlock situation. PE_1 cannot be activated since it is waiting for PE_2 to pass in a data token. PE_2 is in a waiting state since there is no space available in its output to PE_3. PE_3 is in turn waiting for a space to become available in its connection to PE_1.

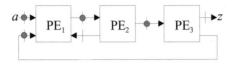

Fig. 12.21 Deadlock wavefront array processor.

In many cases an array processor can be made deadlock free by modifying the queue sizes. For example, Fig. 12.22 shows that the above deadlock situation is eliminated by simply adding a buffer to the connection between PE_2 and PE_3. The operation of the array processor is shown assuming each PE has a unity propagation delay. It is shown that PE_1 accepts an external input a and PE_3 provides an external output z. We assume that the input a queue is filled up as soon as it is vacant. Also, the token placed in output z queue is removed immediately. This ensures that the external I/O will not become a bottleneck of its operation. Note that when $t = 3$, the array processor returns to its original state when $t = 0$. The array processor thus has a cycle time of 3. We will discuss the concept of cycle time in more detail.

The above example brings up the second issue in the design of a wavefront array processor, which is the determination of minimum queue sizes to support a desired operation.

If we study the array processor Fig. 12.22 more carefully and recall that every PE has a delay time of one, we see that the data token placed at input a is consumed every 3 time units. In other words, the maximum input rate to the array processor is one data token every 3 time units. On the other hand, consider the output z; it produces a data token every 3 time units. The throughput of this array processor is thus 1/3, i.e., one result every 3 time units. In Fig. 12.21, when the array processor is in a deadlock situation, its cycle time is ∞ and its throughput is 0. The difference is all in one additional space on one interconnection. It is evident the performance of an array processor depends significantly on the queue sizes. We thus propose another design issue:

- Given queues with infinite capacities, what is the best throughput achievable with an array processor?
- Given a desired achievable throughput (e.g., the best throughput achievable with infinite queues), what is the smallest queue capacity that can support it?

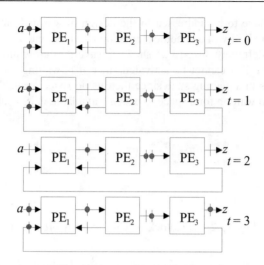

Fig. 12.22 Deadlock free wavefront array processor.

The simple wavefront array structure shown in Fig. 12.23 is used to study these issues. We assume that the array operation is not bounded by its I/O, which implies that input a and output z are always ready. The analysis in Fig. 12.23 shows that this array processor has a cycle time of 3 and a throughput rate (at z) of 1/3. We will discuss a way to determine the best throughput of a design in a moment. Intuitively, the maximum throughput rate achievable by the array processor is 1, which implies a datum is consumed every clock cycle at a.

In Fig. 12.24, we have added one buffer to the array processor. An analysis of its operation determines that it has a cycle time of 2 and thus a throughput rate of 1/2. Fig. 12.25 shows another modification. After the initial latency, the array processor has a cycle time of 1 and a throughput rate of 1.

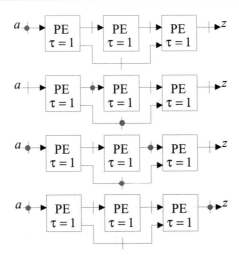

Fig. 12.23 Array processor with a throughput of 1/3.

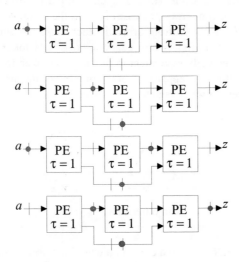

Fig. 12.24 Modified array processor with a throughput of 1/2.

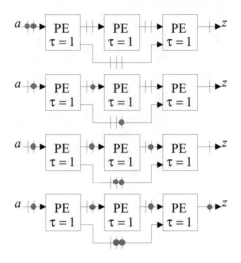

Fig. 12.25 Array processor with a throughput of 1.

Consider a directed cycle (i.e., a cycle with arcs of consistent directions) extracted from a wavefront array processor such as the example shown in Fig. 12.26. Recall that each PE in a wavefront array, in each step, consumes exactly one data token in every one of its inputs and generates one data token at every one of its outputs. This is conceptually equivalent to moving one token from a PE's input to its output at each step. It is thus apparent that the number of tokens in a directed cycle of a wavefront array remains a constant.

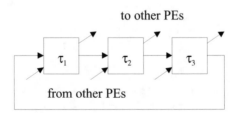

Fig. 12.26 Directed cycle in a wavefront array processor.

Assume that a deadlock situation does not exist. If there is a single data token in the cycle, it will move from the input of a PE to its output and continue to travel along the cycle. A trip around the cycle will take at least $\tau_c = \sum_{i \in c} \tau_i$ time units. In the example of Fig. 12.26, the lower bound of travel time around the cycle is $\tau_c = \tau_1 + \tau_2 + \tau_3$. The pipelining period $T(c)$ of cycle c is thus τ_c in this case. The average throughput of cycle c is $1/\tau_c$.

If there are two data tokens in the directed cycle, the pipelining period is reduced to $\tau_c/2$ but this value cannot be smaller than the longest delay time of a PE. Therefore, the upper bound of the pipelining period of a directed cycle $T(c)$ is $\tau_c/D(c)$, where $D(c)$ is the number of tokens in the cycle.

For a given directed cycle, if we keep on adding tokens to it, its pipelining period will be initially reduced as we described above. However, eventually the cycle operation becomes space dominated. Consider a directed cycle with only one free space and all other spaces are filled with tokens. An analysis of this cycle will show that its pipelining period will once again be bounded by $\tau_c/S(c)$, where $S(c)$ is the number of spaces in the cycle.

Summarizing what we have observed so far we see that

- The number of tokens in any directed cycle $D(c)$ is a constant.
- Similarly, the number of spaces in any directed cycle $S(c)$ remains a constant.
- The pipelining period of any directed cycle $T(c) \geq \tau_i$, where τ_i is the propagation delay of PE_i in the cycle.
- The pipelining period of any directed cycle $T(c) \geq \tau(c)/D(c)$.
- The pipelining period of any directed cycle $T(c) \geq \tau(c)/S(c)$.

While the queue capacity of a wavefront array is set by the designer, the data tokens are decided by the intended operations. The optimal achievable pipelining period of a given wavefront array is thus determined by considering the data tokens:

$$T_{\text{optimal}} = \max_{i,c} \left[\tau_i, \frac{\tau(c)}{D(c)} \right] \tag{12.20}$$

where c is any elementary cycle in the array processor. The pipelining period of an array processor can be only as short as its slowest cycle, so only elementary directed cycles are considered in (12.20). This period can be achieved by providing all queues with infinite capacities. This is of course not a practical solution. We are more interested in finding a design with finite queue capacities to support a given achievable pipelining period.

The optimal pipelining period achievable with a given wavefront array processor can be determined by creating a structure with infinite capacity queues and has an equivalent pipelining period as the original array processor. Both the data tokens and spaces in the original array processor are represented by data tokens in its equivalent structure. The equivalent structure is thus easier to analyze since its pipelining period is determined only by its data tokens.

The equivalent structure of a given array processor can be generated easily as follows. For each arc that goes from PE_i to PE_j, we replace the original arc with one that has an infinite capacity queue and the same number of tokens. In addition, a new arc with an infinite capacity queue that goes in the reverse direction from PE_j to PE_i is added with the number of initial data tokens on it equal to the number of spaces in the original arc. Every arc in the original structure is converted this way into two arcs that form a directed cycle in the new structure.

We now show that an equivalent structure created as described has the same pipelining period as its original wavefront array. This is based on the observation that the character of a wavefront array is determined by its distribution of tokens and spaces. Consider an undirected cycle y of a wavefront array processor. We have the following relationship that combines tokens and spaces into a single property:

$$D_c^t(y) + S_{cc}^t(y) = D_c(y) + S_{cc}(y) = DS_c(y)$$
$$D_{cc}^t(y) + S_c^t(y) = D_{cc}(y) + S_c(y) = DS_{cc}(y)$$

(12.21)

where $D_c^t(y)$ is the number of tokens on clockwise arcs of y at any time t, $D_{cc}^t(y)$ is the number of tokens on counter-clockwise arcs of y at any time t, $S_{cc}^t(y)$ is the number of spaces on clockwise arcs of y at any time t, $S_{cc}^t(y)$ is the number of spaces on counter-clockwise arcs of y at any time t, and $D_c(y)$, $D_{cc}(y)$, $S_c(y)$, and $S_{cc}(y)$ are initial values.

Fig. 12.27 shows an example undirected cycle, in which (12.21) is evaluated to be $DS_c(y) = 3 + 1 = 4$ and $1 + 2 = 3$. After PE$_3$ has been activated, the values become $DS_c(y) = 2 + 2 = 4$ and $DS_{cc}(y) = 0 + 3 = 3$.

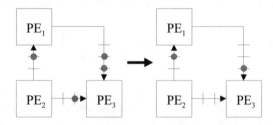

Fig. 12.27 An example demonstrating the undirected cycle property.

Now consider the equivalent structure of a wavefront array. The spaces in an arc (e.g., clockwise) of the original array processor are converted into the data tokens in an arc of opposite direction (e.g., counter-clockwise). An undirected cycle y in the original array processor, as defined in (12.21), is now represented as a directed cycle. Since the number of tokens in a directed cycle is a constant, the relationship described in (12.21) is maintained in the equivalent structure of the array processor. The equivalent structure also maintains the property of the original array processor that the number of tokens and spaces always add up to a constant on an arc. Based on the relationship between the original and new structures, we conclude that the pipelining period of the original structure can be determined by considering the data tokens in the equivalent structure.

The advantage of the equivalent structure is now clear. Only data tokens have to be considered in an analysis. Fig. 12.28 shows an example of converting a wavefront array into its equivalent structure. Note that in the equivalent structure, we do not need to specify the spaces of a queue since there is always an infinite number of spaces available.

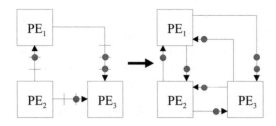

Fig. 12.28 Converting an array processor for pipelining period analysis.

The pipelining period in an equivalent structure is not constrained by the number of spaces available. We apply the relationship between PE delay time and pipelining period to an equivalent structure. The pipelining period T of an equivalent structure is

$$T = \max_{i,c}\left[\tau_i, \frac{\tau_c}{D_c}\right] \tag{12.22}$$

where c is an elementary directed cycle of the equivalent structure, $\tau(c)$ is the sum of delay times in cycle c, and D_c is the number of tokens in the cycle c.

Applying (12.22) to the original DFG, we determine that the pipelining period is

$$T = \max_{i,a,y}\left[\tau_i, \frac{\tau_j + \tau_k}{Q_a}, \frac{\tau(y)}{DS_c(y)}, \frac{\tau(y)}{DS_{cc}(y)}\right] \tag{12.23}$$

where τ_i is the propagation delay of PE_i, a is the arc connecting PE_j and PE_k, Q_a is the queue capacity (sum of token numbers and spaces) of arc a, and $DS_c(y)$ and $DS_{cc}(y)$ are as defined in (12.21).

The inclusion of the first term in (12.23) is trivial since the pipelining period cannot be shorter than the delay of the slowest PE. The second term comes from the fact that each arc a is converted into an elementary directed cycle in the equivalent structure. The number of tokens in this directed cycle is equal to capacity Q_a. The last two terms in (12.23) can be derived by considering that each undirected cycle y in the original structure is converted into two directed cycles c_1 and c_2 in its equivalent structure. The following relations hold for this conversion:

$$D(c_1) = DS_c(y)$$
$$D(c_2) = DS_{cc}(y)$$

(12.24)

An equivalent structure of a wavefront array also provides a means to detect the possibility of having a deadlock in the array processor. Recall that a deadlock situation occurs when two PEs are waiting for each other's data. This will result in a complete halt of the array processing. In other words, the pipelining period $T = \infty$. It is not difficult to see from (12.23) that a finite T exists if and only if $DS_c(y)$ and $DS_{cc}(y)$ are non-zero values. This implies that the equivalent structure of a deadlock free wavefront array must have no empty cycles.

For a given pipelining period T, the following guidelines developed from (12.23) can be used to determine the queue capacity requirements:

$$S_c(y) \geq \frac{\tau(y)}{T} - D_{cc}(y)$$

$$S_{cc}(y) \geq \frac{\tau(y)}{T} - D_c(y)$$

(12.25)

$$S_a \geq \frac{\tau_i + \tau_j}{T} - D_a$$

$$S_a \geq \max(1 - D_a, 0)$$

for any elementary undirected cycle c and any arc a in the wavefront array.

In order to design a wavefront array to support a desired pipelining period, we can apply these guidelines to determine its queue capacities. However, in most cases manual methods are only effective for small problems. For the design of complex wavefront array processors, integer or linear programming can be applied to minimize the total queue capacities while maintaining the desired pipelining period.

12.4 To Probe Further

Array Processors:

- S. Y. Kung, *VLSI Array Processors*, Prentice-Hall, 1988.

- C. Mead and L. Conway, *Introduction to VLSI Systems*, Addison-Wesley, 1980, pp. 271-292.

Systolic Arrays:

- J. McCanny, J. McWhirter, and E. Swartzlander Jr., ed., *Systolic Array Processors*, Prentice-Hall, 1989.

Wavefront Arrays:

*   S. Y. Kung, S. C. Lo, S. N. Jean, and J. N. Hwang, "Wavefront array processor–concept to implementation," *IEEE Computer*, July 1987, pp. 18-33.

Systolic Array Application:

*   C. L. Wang, C. H. Wei, and S. H. Chen, "Efficient Bit-Level Systolic Array Implementation of FIR and IIR Digital Filters," *IEEE Journal on Selected Areas in Communications, Vol. 6, No. 3*, April 1988.

Linear Programming:

*   W. H. Press, S. A. Teukolsky, W. T. Vetterling, and B. P. Flannery, *Numerical Recipes in C : The Art of Scientific Computing, 2nd Edition*, Cambridge University Press, 1993.

Problems

12.1   Explore the valid combinations of the hyperplane set and projection directions in Fig. 12.14. Compare and determine the best array processing architecture for the matrix-vector multiplication problem.

12.2   Design an array processor to multiply two 2×2 matrices and produce a 2×2 product matrix.

Chapter 13   Fault Tolerant VLSI Architectures

Be Prepared ...

It is evident from our discussion in Chapter 12 that one of the intentions of creating application specific VLSI architectures is to achieve high performance by using efficiently as many PEs (processing elements) as possible. Ideally, we would like to implement all PEs of a system on the same chip to minimize the interconnection lengths.

The number of components that can be reliably fabricated on a chip is limited by the fabrication process. Wafer scale integration (WSI) was proposed in the 1980's and received a lot of enthusiasm initially. WSI intends to create and connect all chips in a system on the same wafer. Unfortunately, the yield of a fabrication process is inversely proportional to the chip (wafer) area. It was soon clear that the low yield of WSI limited its practical applications.

The WSI concept has been replaced by the multi-chip module (MCM) technology. The MCM approach overcomes the packaging limits of printed circuit boards (PCB) by eliminating chip packages. Instead, chips are mounted directly on a silicon wafer with layers of metal interconnections. Chips are typically an order of magnitude smaller than package areas. Without the packages, chips can be placed closer on the silicon wafer for higher density and faster interconnections. The elimination of packaging also significantly reduces the parasitic capacitance and inductance which have been the major obstacles of high performance.

Regardless of the approach used to assemble a large number of PEs in a system, reliability remains an important factor for its success. A normal system requires all PEs to be functional.

Fig. 13.1 shows that the reliability of a system goes down exponentially with the number of PEs. Assume each PE has a reliability of 0.999. The overall reliability of a 1000 processing element systolic array is merely $0.999^{1000} = 0.37$.

Fault tolerance is a potential solution to a system with a large number of PEs. Before we discuss the construction and operation of fault tolerant VLSI architectures, an introduction is given to reliability and fault tolerance techniques. This introduction is necessarily brief and simplified.

13.1 Reliability

In the following discussion, we will treat a PE as the smallest building block in a VLSI system. Please note that the derivation of reliability applies equally to other levels of building blocks (e.g., gates, macro-cells, etc.). A fault-tolerant VLSI system is defined as one that can continue to provide its intended service even after certain PEs have failed. We have discussed techniques to test VLSI circuits in Chapter 9. In fabrication line testing, a fault tolerant system allows a faulty circuit to be repaired. A working system with on-line fault detection can use its fault tolerant capability to provide adequate service in spite of the presence of a fault. Based on these principles, there are two aspects of fault tolerance. Techniques have been developed to tolerate failures in the field. Approaches have also been evolved to tolerate manufacturing defects to achieve a better yield rate.

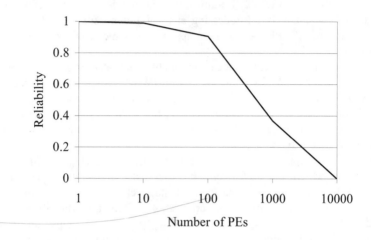

Fig. 13.1 Reliability of a system with multiple PEs.

From the viewpoint of fault tolerance, PEs can be classified as good, bad, or marginal. A marginal PE functions as intended most of the time, but not all of the time. Only good PEs can be accepted as being reliable.

Reliability is defined as the probability that a device will operate adequately for a given period in its intended application. Consider that n identical PEs are produced. Let $s(t)$ be the number of PEs that are still functioning. The number of failed PEs up to time t is thus $n - s(t)$. The probability of a PE being functional at time t is defined as the reliability $r(t)$, and

$$r(t) = \frac{s(t)}{n} \qquad (13.1)$$

Another useful parameter is the failure rate $h(t)$ of a PE, which is defined as the number of failures per unit time in the group of surviving PEs,

$$h(t) = \frac{1}{s(t)} \times \frac{d(n - s(t))}{dt} \qquad (13.2)$$

A typical failure rate $h(t)$ curve is shown in Fig. 13.2, which is often referred to as a bathtub curve for an obvious reason.

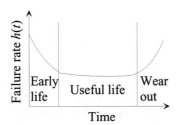

Fig. 13.2 A typical failure rate curve.

Components that have passed the initial quality control test but do not function as intended after being put in operation contribute to failures in the early life period. A burn-in process, in which a system is operated at an elevated stress level (e.g., raised temperature, voltage, etc.), can be used to screen out early failures. In the useful life period of a system, the failure rate $h(t)$ can be approximated as a constant. The reliability of a PE with a constant failure rate $h(t) = \eta$ is derived as follows.

$$r(t) = \frac{s(t)}{n} = 1 - \frac{n - s(t)}{n},$$
$$\frac{dr(t)}{dt} = -\frac{1}{n} \times \frac{d(n - s(t))}{dt} \tag{13.3}$$

$$\eta = \frac{1}{s(t)} \times \frac{d(n - s(t))}{dt} = -\frac{n}{s(t)} \times \frac{dr(t)}{dt} = -\frac{1}{r(t)} \times \frac{dr(t)}{dt} \tag{13.4}$$

Solving (13.4), we have

$$r(t) = e^{-\eta t} \tag{13.5}$$

We refer to (13.5) as the exponential failure law. The constant failure rate η is usually expressed as the percentage of PEs failed in a period of time. If ηt is small, we can simplify (13.5) into

$$r(t) = 1 - \eta t \tag{13.6}$$

For a system with k different types of PEs, each of which has a failure rate of η_i, the overall system failure rate η_v is

$$\eta_v = \sum_{i=1}^{k} n_i \eta_i \qquad (13.7)$$

where n_i is the number of PEs in each type.

Since the probability that a system will perform successfully depends upon the conditions under which it operates, the reliability figure is not necessarily a practical measure for system evaluation. A more useful parameter is the average time that a system will operate before a failure stops it from delivering services. This parameter is called the mean-time-before-failure (MTBF).[1]

$$\text{MTBF} = \int_{t=0}^{\infty} r(t)dt \qquad (13.8)$$

For the exponential failure law, substitute (13.5) into (13.8),

$$\text{MTBF} = \int_{t=0}^{\infty} e^{-\eta t}dt = \frac{1}{\eta} \qquad (13.9)$$

If we use (13.6) in (13.8),

$$\text{MTBF} = \frac{t}{1 - r(t)} \qquad (13.10)$$

Another useful parameter in the study of fault tolerance is maintainability, which is defined as the probability of detecting and locating a fault and repairing it in a given time. A term similar to MTBF can be defined for the maintenance action, which is the mean-time-to-repair (MTTR). If MTTR is reduced, maintainability is increased. In order to design and manufacture a maintainable system, it is important to predict the MTTR for various fault conditions that could occur in the system. The expertise of designers and the circuit testability are invaluable for the improvement of MTTR. The system repair time, which is mainly spent on the detection and isolation of faults, can be improved significantly by a design-for-testability approach.

The availability of a system is defined as the probability that a system will be functional:

---

[1] MTBF is also the acronym for mean-time-between-failures, which means roughly the same thing.

$$\text{Availability} = \frac{\text{System up-time}}{\text{System up-time} + \text{System down-time}}$$

$$= \frac{\text{MTBF}}{\text{MTBF} + \text{MTTR}}$$

(13.11)

13.2 Fault Tolerant Design

The key of fault tolerant systems is redundancy. Redundancy means that two or more identical copies of the same PEs are employed. We describe a number of popular redundancy schemes in the following sections. These redundancy systems were developed for general computing systems so they may not be directly applicable to VLSI architectures. However, this brief introduction should provide the necessary background for our further discussion of fault tolerant VLSI systems.

Redundancy is a way of masking errors; so it is also called masking redundancy. This technique uses extra PEs in the system so that the effect of a faulty component is masked instantaneously. Even though a duplicated system (duplex) theoretically doubles the reliability of an original simplex system, its operation depends on the availability of a method to detect the faulty PE. For example, assume PE_1 in a duplex system produces the output of 1 and PE_2 in the same system produces the output of 0. It cannot be determined which PE is right without applying further fault detection.

Providing a third redundant PE to a duplex system makes a triple modular redundancy (TMR) system. A TMR system is the most popular hardware masking technique. An illustration of a TMR system is shown in Fig. 13.3.

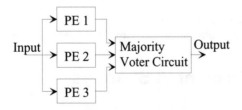

Fig. 13.3 A TMR system.

A TMR system can tolerate the failure of one PE. Assuming that only one fault can occur at a time (i.e., single fault assumption), a TMR system has the advantage of being able to perform immediate fault-masking actions. The majority voter will select the output of the two fault-free PEs and send it to the system output. Both temporary and permanent faults are covered by a TMR system. In addition to being quite straightforward in its implementation, the TMR system does not require a fault detection to be performed before masking.

The following discussion assumes that the majority voter circuit is permanently fault-free (i.e., perfect). This assumption is often justified by the fact that the majority voter is substantially simpler than the PEs. The overall reliability of the TMR system in Fig. 13.3 is

$$r_v(t) = r^3(t) + 3r^2(t)(1 - r(t)) = 3r^2(t) - 2r^3(t) \qquad (13.12)$$

A little inspection of (13.12) should reveal that it is the standard probability equation for having 2 or more PEs being functional in the system. This is a direct result of the reliability definition.

If the majority voter has a less than one reliability, then the reliability of a TMR system has to include its reliability $r_{voter}(t)$, and

$$r_v(t) = r_{voter}(3r^2(t) - 2r^3(t)) \qquad (13.13)$$

The above equation demonstrates the fact that the reliability of a system is limited by the component with the lowest reliability. In other words, a system can only be as reliable as its least reliable component.

The TMR concept can be expanded to an NMR (N-modular redundancy) approach. In general, an NMR system can tolerate up to m module failures, and $m = (N-1)/2$. In practice, N is usually an odd number for an obvious reason. If the voter is perfectly reliable, then

$$r_v(t) = \sum_{i=0}^{n} \binom{N}{i} (1 - r(t))^i r^{N-i}(t) \qquad (13.14)$$

Other redundancy schemes have been developed such as dynamic redundancy which uses the output of one module at a time and treats the other modules as spares.

13.3 Fault Tolerant VLSI Arrays

A single VLSI structure may contain hundreds or thousands of cells connected in a more or less near-neighbor array structure. These cells could be memories, PEs, or any other functional blocks. While such structures have many potentials, the failure of a single cell in them disrupts their service.

Redundancy can be used to alleviate this problem. The design can be done in a way to allow the bypassing of faulty cells within the array. The fault-free cells are interconnected to achieve a functional array.

When spare cells are provided, a failed cell can be switched out and replaced by a spare. This is called the reconfiguration of an array. A successful reconfiguration will allow the array to operate even if a certain number of faults are present.

An array may contain failed cells as soon as it comes off the production line. A fabrication reconfiguration can be performed immediately after manufacturing to improve the yield. Many applications (e.g., satellite operations) will require that the array remain operational in spite of failed cells. In some cases, the array is allowed to shut down to perform an off-line reconfiguration, while in other cases a real time reconfiguration is performed as the array continues to operate.

Our discussion below is targeted toward array processors of the type discussed in Chapter 12; however, the reconfiguration algorithms are readily adaptable to other VLSI structures.

A reconfiguration scheme must specify how the spare PEs are incorporated with the original array. In the event of a fault, the way that a functional array is formed by replacing faulty PEs with spare PEs must also be described. VLSI arrays can be designed in many ways to incorporate spare PEs. However, the most popular method to provide redundancy is by means of spare rows or spare columns or both. Therefore, we specify a fault tolerant processor array by its size as $(m + r) \times (n + c)$, where m and n are the number of rows and columns required to perform the functional operation, respectively, while r and c are the number of spare rows and spare columns, respectively.

In a fault tolerant array structure, a fault-free PE is called an available PE. In contrast, an unavailable PE is a faulty one or a fault-free one that has already been used to replace another PE.

When a fault is detected, a reconfiguration algorithm specifies a way of mapping a logical array of PEs (i.e., the array processor intended by its designer) into a physical array of PEs. Given a PE in a logical array, the reconfiguration algorithm determines which physical PE should be used to carry out its operations. This mapping must be dynamically done according to the availability of spare PEs.

13.4 Set Replacement Algorithm

The set replacement algorithm treats each row of PEs as a set. An array that implements the set replacement algorithm augments an $n \times m$ array with m spare rows. The result is an $(n+r) \times m$ array. In the set replacement algorithm, a row that contains one or more faulty PE is skipped. In other words, a single faulty PE will cause an entire row of PEs to be replaced.

As an example of the set replacement algorithm, consider the array shown in Fig. 13.4. The original 4×4 array has been augmented with one spare row. Each PE is labeled using the notation (i, j) to represent the PE in the ith row and the jth column. Such an array can be scanned top down row by row to configure a logical array. If a row contains a faulty PE, then the row is bypassed and replaced by the next row. When row i is the faulty set, the PEs $(i, 1)$, $(i, 2)$, $(i, 3)$, and $(i, 4)$ are replaced by PEs $(i+1, 1)$, $(i+1, 2)$, $(i+1, 3)$, and $(i+1, 4)$, respectively. Apparently, this array can tolerate up to 4 faulty PEs, as long as they are all in the same row. Fig. 13.5 illustrates the result of applying the set replacement algorithm.

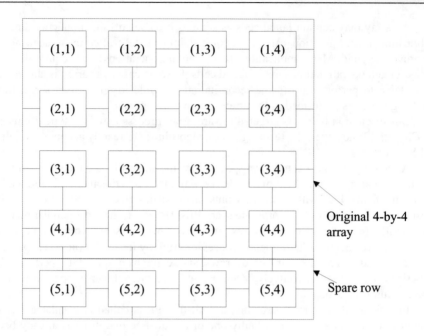

Fig. 13.4 5 × 4 fault tolerant array processor.

Fig. 13.6 shows the use of a multiplexer to skip PE (i, j). Note that this reconfigurable routing structure assumes that the signal flows from the top to the bottom. In the set replacement algorithm, all multiplexers in the same row are set by the same control signal so that the entire row would be bypassed in the case of a failed PE. Additional routing structures have to be provided for the first and last columns to accommodate the routing of horizontal signals (see Problem 13.3).

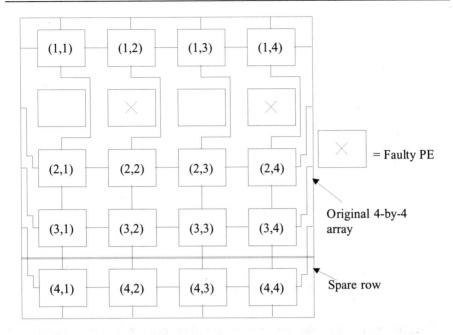

Fig. 13.5 Array reconfigured using the set replacement algorithm.

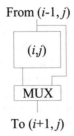

Fig. 13.6 Reconfigurable routing structure for set replacement algorithm.

A variation of the above algorithm provides both spare columns and rows within the array. The array map in Fig. 13.7 shows a $(6 + 2) \times (6 + 2)$ fault tolerant array. There are 6 fault PEs. The success of a reconfiguration depends on the order of replacement. For example, suppose we use the spare rows to replace rows 1 and 3. We will need three spare columns to complete the reconfiguration. The reconfiguration thus fails since only two spare columns are provided. Alternatively, if we use the spare rows to replace rows 3 and 5, then the spare columns can be used to replace columns 2 and 4.

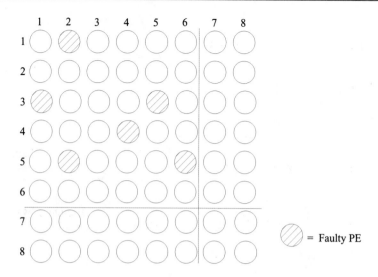

Fig. 13.7 A $(6 + 2) \times (6 + 2)$ fault tolerant array.

A replacement algorithm has been developed to choose spare rows/columns with the guidance of a bipartite graph. A graph contains a set of nodes and a set of edges. Each edge connects two nodes. In a bipartite graph, the set of nodes is divided into two subsets. Each edge has one node in one subset and another node in the other subset.

The set replacement array reconfiguration problem can be represented graphically with a bipartite graph. The corresponding graph for the array in Fig. 13.7 is shown in Fig. 13.8. The nodes on the left hand side of the graph represent the rows that contain faulty PEs. The nodes on the right hand side of the graph represent the columns that contain faulty PEs. For example, node R1 represents row 1 and node C1 represents column 1. An edge is provided between a row and a column node if a faulty PE is located in their intersection. In other words, if the PE at location (i, j) is faulty, then nodes Ri and Cj are connected with an edge. For example, nodes R1 and C2 are connected to indicate that there is a faulty PE in location $(1, 2)$.

The replacement problem is to eliminate all connections between nodes in the bipartite graph. For example, if we choose to replace row 3 with a spare row, the faulty PEs $(3, 1)$ and $(3, 5)$ are replaced. Thus the connections between nodes R3 and C1 and the connection between nodes R3 and C5 are eliminated.

The node that has the largest number of edges connected to other nodes is identified and replaced. After a node is replaced, the node and its edges are removed from the graph. This process is repeated until either the array is successfully reconfigured or all the spares are used up. In our example, we replace row 3, row 5, column 4, and column 6.

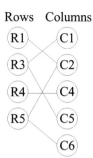

Rows   Columns

Fig. 13.8 Bipartite graph for reconfiguring the array in Fig. 13.7.

Since the above algorithm pursues to use a set to replace a row or column with the largest number of faulty PEs first, it is a greedy algorithm. As in the case of any greedy algorithm, it may have to deal with the possibility of being trapped in a local minimum.

The replacement of a set of PEs at a time simplifies the control structure, but it does not allow an efficient use of the spare PEs. We discuss a structure that utilizes the spare PEs more efficiently below.

13.5 Shifting Replacement Algorithm

The shifting replacement algorithm is a variation of the set replacement algorithm. The routing structure for each PE is individually configured. Due to the additional complexity for the routing structure, typically only a single row of spare PEs is added to the original array. The algorithm to reconfigure the array is to shift up the PEs below the one to be replaced. The functions of the faulty PE are performed by its lower neighbor, of which the functions must be transferred to the next neighbor PE. This process continues until the PE from the spare row is used.

Fig. 13.9 shows the result of performing the shifting replacement algorithm in the array shown in Fig. 13.4. The shifting replacement algorithm can handle up to 4 faulty PEs in this array, as long as there is at most one faulty PE in each column. A reconfigurable routing structure capable of supporting the shifting replacement algorithm is shown in Fig. 13.10. This routing structure assumes that the signal in the array flows vertically from top to bottom and horizontally from left to right.

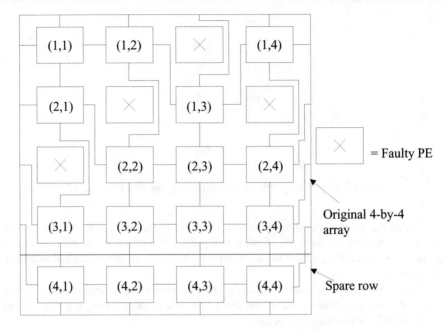

Fig. 13.9 Shifting replacement algorithm.

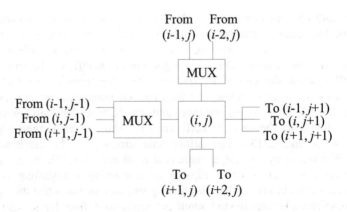

Fig. 13.10 Routing structure for the shifting replacement algorithm.

13.6 Fault Stealing Replacement Algorithm

Neither the set replacement algorithm nor the shifting replacement algorithm can accommodate more than one faulty PE in any one column of an array. In theory, multiple spare rows can be provided for this purpose. However, the complexity of the routing structure will be increased significantly. The fault stealing replacement algorithm provides coverage for multiple fault problems by using a spare column and a spare row of PEs.

In the fault stealing approach, an $(n+1) \times (n+1)$ physical array is used to support an $n \times n$ logical array. Fig. 13.11 shows a $(4+1) \times (4+1)$ array that supports the fault stealing replacement algorithm.

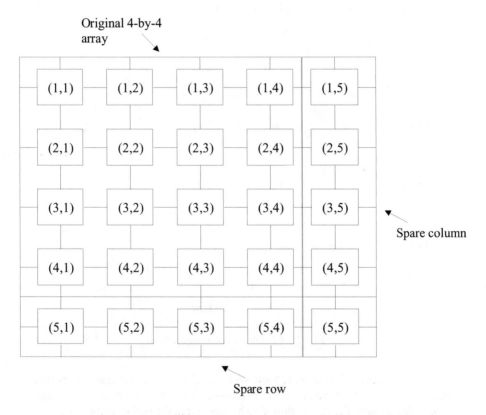

Fig. 13.11 4×4 array augmented with one spare column and one spare row.

The basic fault stealing reconfiguration algorithm is called the direct reconfiguration. The general idea of this algorithm is to scan each column of the array from bottom to top. If a faulty PE is found in a column, it is marked as a vertical fault and the scanning for that column is terminated. After all columns have been scanned, the remaining unmarked faulty PEs are marked as horizontal faults.

Once the faulty PEs are marked as horizontal and vertical faults, the array is physically reconfigured. Horizontal faults are shifted right, and vertical faults are shifted down. An example illustrating this direct reconfiguration is shown in Fig. 13.12. Horizontal faults are marked with "H" and vertical faults are marked with "V". The PEs in the result array are labeled with their locations in the logical array.

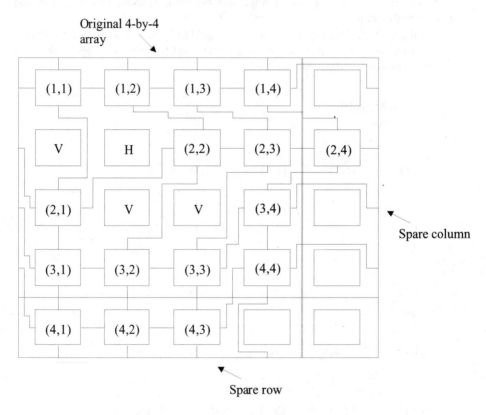

Fig. 13.12 Direct reconfiguration.

The direct reconfiguration scheme will fail if there is more than one horizontal fault in any row of the array. Many extensions have been proposed to change the domain that a faulty PE can be shifted to by modifying the replacement rules.

One of these modified stealing algorithms is called the complex stealing algorithm. Assume that there is one spare row and one spare column. Consider a row that has a number of unavailable (i.e., faulty or stolen) PEs. For an unavailable PE (i, j), the first consideration is to shift it down to $(i+1, j)$ if the PE at that location is available. If it is not, then the stealing of the PE at location $(i+1, j+1)$ is considered. If this is still not possible, then the operations of the unavailable PE are shifted right to location $(i, j+1)$. In order to more efficiently utilize the spare PEs, the last unavailable PE in a row is always shifted right to use the spare PE. An example of applying the complex fault stealing scheme to reconfigure an array is shown in Fig. 13.13. Since there would have been two horizontal faults in the first row of the array, the direct reconfiguration scheme cannot deal with this situation.

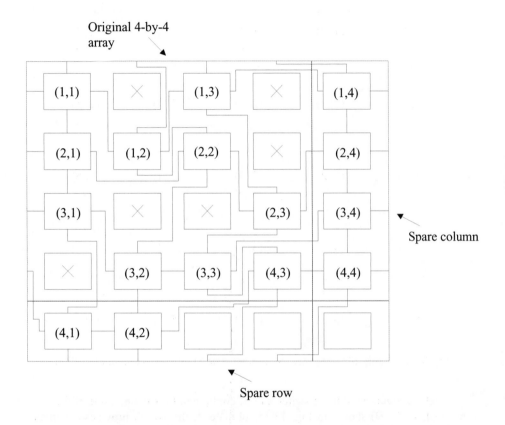

Fig. 13.13 Complex fault stealing algorithm.

While the complex fault stealing algorithm can achieve a high utilization rate of spare PEs, its implementation requires a very complex routing structure, which is evident from the density of routing channels in Fig. 13.13. In fact, the development of a reconfiguration algorithm must balance its replacement flexibility and routing complexity. A higher flexibility in a replacement scheme generally implies a more complex routing structure. Next, we introduce a compensation path replacement algorithm that requires a relatively simple routing structure.

13.7 Compensation Path Replacement Algorithm

A fault tolerant array based on the compensation path replacement algorithm has one spare row and one spare column. Fig. 13.14 shows a $(4 + 1) \times (4 + 1)$ array structured for the compensation path replacement algorithm. The configurable routing switches are indicated with circles in Fig. 13.14.

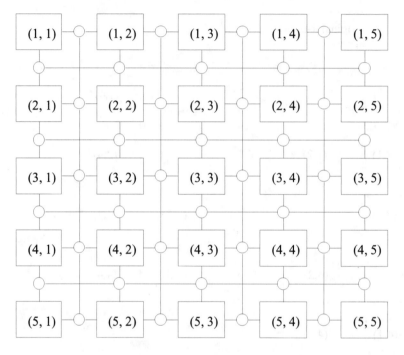

Fig. 13.14 Fault tolerant array for compensation path replacement algorithm.

Each configurable routing switch can be configured to operate in one of the four states (A, B, C, D) shown in Fig. 13.15. In state A, the switch provides a vertical connection. In state B, the switch provides a horizontal interconnection. States C and D provide two different ways to change routing directions. These routing switches are called single track switches since they provide only one communication path in each direction of switching.

Fig. 13.15 Four states of the routing switches used in Fig. 13.14.

Each PE is provided with both a vertical feedthrough channel and a horizontal feedthrough channel. These feedthrough channels are assumed to be fault-free. They allow a faulty PE to be converted into either a vertical or a horizontal interconnecting element. Fig. 13.16 shows the use of these switches to perform a reconfiguration.

The replacement of faulty PEs in this array is performed in a way similar to the shifting replacement strategy. The path of shifting the PEs is called a compensation path. Any replacement is feasible with the routing structure as long as the compensation paths are straight and compensation paths do not intersect.

Fig. 13.16 shows the result of a reconfiguration using compensation paths. A logical 4 × 4 array is configured from a 5 × 5 physical array. The compensation paths are illustrated by the bold face arrows.

13.8 Summary

After a brief introduction of reliability and conventional fault tolerant techniques, we discussed the reconfiguration of VLSI array structures for the purpose of fault tolerance. A set replacement algorithm is easy to implement and simple to control, but it results in a low utilization rate of spare PEs. The development of a reconfiguration strategy must balance between the efficiency of spare cell utilization and the complexity of the reconfiguration structure. After discussing a number of individual replacement algorithms, a reconfiguration structure based on compensation paths was introduced.

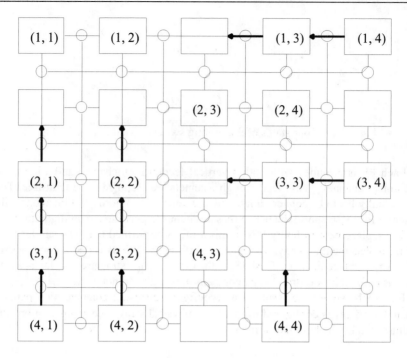

Fig. 13.16 Reconfiguration using compensation paths.

13.9 To Probe Further

Fault Tolerant Systems:

- B. W. Johnson, *Design and Analysis of Fault Tolerant Digital Systems*, Addison-Wesley, 1989.

- V. P. Nelson, "Fault-tolerant computing: fundamental concepts," *IEEE Computer*, July 1990, pp. 19-25.

Fault Tolerant VLSI Systems:

- B. W. Johnson, *Design and Analysis of Fault Tolerant Digital Systems*, Chapter 6, Addison-Wesley, 1989.

- I. Koren and A. D. Singh, "Fault tolerance in VLSI circuits," *IEEE Computer*, July 1990, pp. 73-83.

- J. A. Abraham, P. Banerjee, C-Y. Chen, W. K. Fuchs, S-Y. Kuo, and A. L. N. Reddy, "Fault tolerance techniques for systolic arrays," *IEEE Computer*, July 1987, pp. 65-74.

13.10 Problems

13.1 Consider a redundant system formed of n identical modules. The reliability of each model is $R = e^{-at}$, where a is the constant failure rate. Calculate the overall system reliability and MTBF for the following cases:
(a) The system can tolerate at most 1 faulty module.
(b) The system can tolerate at most $(n - 1)$ fault modules.
(c) The system can tolerate at most $(n - r)$ fault modules.

13.2 The MTBF for a PE with a failure rate of a is $1/a$. Calculate the MTBF for a TMR system that contains three PEs. Compare their MTBFs.

13.3 Design the configurable routing structure for the PEs in the first and fourth columns of the set replacement fault tolerant array shown in Fig. 13.4.

13.4 A $(5+1) \times (5+1)$ fault tolerant array with faulty PEs at locations $(1, 2)$, $(1,3)$, $(3, 5)$, $(2, 4)$, $(5, 1)$ is to be reconfigured into a 5×5 array. Show the reconfiguration using the following algorithms:
(a) Set replacement.
(b) Shifting replacement.
(c) Direct fault stealing.
(d) Complex fault stealing.
(e) Compensation path.

INDEX

A

B

C